BIRDS - EVOLUTION, BEHAVIOR AND ECOLOGY

BIRDS

EVOLUTION AND BEHAVIOR, BREEDING STRATEGIES, MIGRATION AND SPREAD OF DISEASE

BIRDS - EVOLUTION, BEHAVIOR AND ECOLOGY

Additional books in this series can be found on Nova's website under the Series tab.

Additional e-books in this series can be found on Nova's website under the e-book tab.

BIRDS - EVOLUTION, BEHAVIOR AND ECOLOGY

BIRDS

EVOLUTION AND BEHAVIOR, BREEDING STRATEGIES, MIGRATION AND SPREAD OF DISEASE

LUCAS RUIZ
AND
FRANCO IGLESIAS
EDITORS

NOVA BIOMEDICAL

New York

For permission to use material from this book please contact us:
Telephone 631-231-7269; Fax 631-231-8175
Web Site: http://www.novapublishers.com

Library of Congress Cataloging-in-Publication Data

Birds : evolution and behavior, breeding strategies, migration and spread of disease/ editors, Lucas Ruiz and Franco Iglesias.
pages cm
Includes bibliographical references and index.
ISBN: 978-1-62948-104-3 (hardcover)
1. Birds--Diseases. 2. Birds--Evolution. 3. Birds--Behavior. 4. Birds--Breeding. I. Ruiz, Lucas. II. Iglesias, Franco.
SF994.B57 2013
598.13'8--dc23

2013034690

Published by Nova Science Publishers, Inc. † New York

Contents

Preface

In this book, the authors present current research in the study of the evolution and behavior, breeding strategies, migration and spread of disease within the bird species. Topics discussed in this compilation include sex determination in birds; structure and function of the avian respiratory system, with discussion regarding its predisposition to injury by particulates and pathogenic microorganisms; distribution and dispersion of coccidia in wild passerines of the Americas; an animal geography of the dominant urban avian scavenger in contrasting case studies; recent infectious diseases or their responsible agents recorded from Japanese wild birds; ticks on Brazilian birds; an overview of recent parasitic diseases due to helminths and arthropods recorded from wild birds, with special reference to conservation medical cases from the Wild Animal Medical Center of Rakuno Gakuen University in Japan; environmental factors that affect urban avian communities; and the impact of landscape configuration and competitors on hooded vulture necrosyrtes monarchus temminck 1823 in Southern Ghana.

Chapter 1 - How gender is formed in birds? What caused dimorphism and formation of biologically important feature? What are the internal physiological mechanisms and the role of the genome in the process? To all these questions are answered offered to the reader head chapter. Paper is devoted to sex determination in birds. It will be reviewed data on the genome of birds, information on the sex chromosomes, heterogametic, sex of birds of the different karyotype. Specificity of Z and W chromosomes, sex chromosomes in Ratinae will be considered. The ideas will be presented about the role of hormones in sex determination and differentiation. The data on the possibility of "inversion" of sex in birds in the inhibition of aromatase, estrogen exposure and gonadectomy will be introduced. In addition, review key sex determining genes, chromosomal determinants, specific dosage compensation will be discussed. Sex determining genes will be examined in birds, including *Dmrt1, AMH, FOXL2* and W chromosomal determinants. The specificity of birds dose compensation will be described. The data will be submitted about MHM region of Z chromosome. Features of sex determination in males and females will be discussed. Bird cell-autonomous sexual identity (CASI) will be considered.

Chapter 2 - Among the air-breathing vertebrates, the avian respiratory system, the lung-air sac system, is structurally the most complex and functionally the most efficient gas exchanger. The lung which is small and compact is firmly held between the ribs and the vertebrae while the air sacs are capacious transparent membranous structures which are widely spread in the coelomic cavity. The lung is ventilated unidirectionally and continuously by a bellows-like synchronized activity of the air sacs. It takes two inspiratory cycles and two exhalatory cycles for the air to enter and exit the lung. Morphometrically, the avian lung is highly specialized. Compared to a mammal of equivalent body mass, the bird lung has a respiratory surface area which is 15% greater and the blood-gas barrier is 62% thinner. Free (surface) macrophages are scarce on the the respiratory surface of the avian lung. Because of this it has been argued that birds have poor pulmonary cellular defences and are therefore highly vulnerable to injury by particulates and infection by pathogens. This conclusion has, however, been questioned. Diverse phagocytic cells which include subepithelial macrophages, pulmonary intravascular macrophages, bronchial epithelial cells, and free macrophages should provide adequate protection. The large tidal volume of birds and the extension of the air sacs to lie superficially, are some of the features which make birds susceptible to pulmonary afflictions and infections.

Chapter 3 - Reports of coccidia of the genera *Eimeria* Schneider, 1875 and *Isospora* Schneider, 1881 from passerine birds have been reported for more than two centuries. In the Americas, the order Passeriformes comprises 47 families containing 2,453 species. *Isospora* species have been observed in 23 (49%) of the 47 families (Dendrocolaptidae, Furnariidae, Thamnophilidae, Cotingidae, Tyrannidae, Corvidae, Meliphagidae, Mimidae, Cardinalidae, Coerebidae, Emberizidae, Estrildidae, Fringillidae, Hirundinidae, Icteridae, Parulidae, Passeridae, Sturnidae, Thraupidae, Timaliidae, Turdidae, Troglodytidae, Zosteropidae) and in 66 (3%) of the 2,453 species. A total of six *Isospora* species have been described from *Ramphocelus bresilius dorsalis* (Sclater); four from *Camarhynchus parvulus* (Gould); and *Saltator similis* (Lafresnaye d'Orbigny); three from each of the following: *Sporophila caerulescens* (Vieillot); *Sporophila frontalis* (Verreaux); *Oryzoborus angolensis* L.; *Thraupis sayaca* L.; and *Zosterops japonicus* (Temminck, Schlegel); two from each of the following: *Dendrocolaptes certhia* (Sclater & Salvin); *Xiphorhynchus ocellatus* (Spix); *Saltator grossus* L.; *Coereba flaveola* L.; *Geospiza fortis* (Gould); *Lonchura punctulata* L.; *Serinus canaria* L.; *Thraupis palmarum* (Wied); *Turdus albicollis* (Vieillot); and *Sicalis flaveola* L.; and a single species from each of the following: *Glyphorynchus spirurus* (Vieillot); *Dendrocincla merula merula* (Lichtenstein); *Dendrocincla merula barletti* (Chubb); *Hyloctistes subulatus* (Spix); *Sclerurus caudacutus* (Vieillot); *Automolus infuscatus* (Sclater); *Hylophylax naevioides* (Lafresnaye); *Procnias nudicollis* (Vieillot); *Myiarchus ferox* (Gmelin); *Mionectes rufiventris* (Cabanis); *Corvus brachyrhynchos* (Brehm); *Calocitta formosa* (Swainson); *Cyanocorax chrysops* (Vieillot); *Foulehaio carunculatus* (Gmelin); *Cardinalis cardinalis* L.; *Paroaria coronata* (Miller); *Tiaris fuliginosus* (Wied); *Zonotrichia capensis* (Müller); *Gubernatrix cristata* (Vieillot); *Carduelis carduelis* L.; *Carduelis atrata* (Lafresnaye d'Orbigny); *Carduelis tristis* L.; *Carduelis chloris* L.; *Fringilla coelebs* L.; *Hemignathus virens* (Cabanis); *Petrochelidon pyrrhonota* (Vieillot); *Dives atroviolaceus* (Lafresnaye d'Orbigny); *Cacicus cela* L.; *Sturnella bellicosa* (Filippi); *Icterus icterus* L.; *Icterus graceannae* (Cassin); *Geothlypis aequinoctialis* (Gmelin); *Passer domesticus* L.; *Gracula religiosa* L.; *Leucopsar rothschildi* (Stresemann); *Chlorospingus ophthalmicus* (Du Bus Gisignies); *Iridosornis analis* (Tschudi); *Dacnis cayana* L.; *Leiothrix lutea* (Scopoli); *Myadestes obscurus* (Gmelin); *Catharus aurantiirostris* (Hartlaub); *Turdus migratorius* L.; *Turdus grayi* (Bonapart); *Turdus assimilis* (Cabanis); *Turdus plebejus* (Cabanis); *Thryophilus rufalbus* (Lafresnaye); *Cantorchilus modestus* (Cabanis); and *Mimus gilvus* (Vieillot). Observations in relation to *Eimeria* species have been significantly less frequent. Specifically, the species *Eimeria divinolimai* Berto, Flausino, Ferreira, Lopes, 2008 and *Eimeria sicki* Berto, Luz, Flausino, Ferreira, Lopes, 2009 were described from two hosts only [*Casiornis rufus* (Vieillot) and *Myiarchus ferox* (Gmelin)] both of which belong to the family Tyrannidae. This chapter contributes to the author's understanding of the distribution of coccidian parasites of passerine birds from Northern, Central and Southern America. To this end, the coccidia were organized and grouped according to the family of the host, emphasizing the geographic location of the reports and associated aspects.

Chapter 4 - Animal geography and avian ecology represent two ways of looking at avian relations with the environment and human behavior. Animal geography is the geography of animals, including the examination of animals as actants; active agents creating relations with people as individuals and groups. Avian ecology is more concerned with statistical generalization, at inter- and intra- species levels. Urban areas provide instructive case studies for the application of such perspectives, due to the intense relation between animals and people in highly variable landscapes. In this chapter, case studies of the application of these paradigms and sub-disciplines are taken from published articles on urban birds in North America, Europe and Africa. In all the studies there were similar issues; dominant scavengers in relationship with smaller birds, and with people acting as non-feeders, and accidental and deliberate feeders. From the perspective of urban avian ecology, dominant avian scavengers obtained a statistically significant dominant relationship with smaller scavengers. From the perspective of animal geography, both dominant and dominated scavengers constructed individual strategies to enable food access. This individualism, in reaction to other birds, and towards variable, individualist human behaviors, could be defined as actancy, with evidence of innovation, experimentation and possibly strategic thinking. These studies illustrate the importance of flexibility in avian ecological studies.

Chapter 5 - Most infectious micro agents, e. g., parasitic and/or commensal viruses, bacteria, fungus and protozoans, are component parts of the ecosystems in which their wild avian hosts occur, they have co-evolved, and the infectious agents do not necessarily cause disease. Sometimes, however, ten of thousands of birds become sick and in extreme cases die, in various parts of the world because of an outbreak of an epidemic disease. By the way, considerable environmental change has occurred in Japan over the last 150 years as well. Furthermore, Japan is an important transit country for a wide range of migratory species. Japan is situated on the East Asian Flyway a migratory route, connecting northeast Asia with Southeast Asia, with important branches passing through the Nansei Shoto, Kyushu, Honshu and Hokkaido into northeast Russia and also via Kyushu and the Korean Peninsula into eastern China. The outbreaks of infectious disease have not so far significantly impacted populations of birds in Japan. Because an understanding of the infectious avian disease organisms that are present in Japan, their ecology, transmission, and symptoms is an essential conservation tool, the authors have overviewed this field by the end of the 20 century. However, according to the decade passed by, several issues have to be added to the review. Hence, this chapter provides a recent overview of the infectious diseases and/or their responsible micro agents recorded in Japan, and briefly potential strategies for epidemic risk reduction are given. For example, in viral diseases or agents, avian pox, rotaviral enteritis or arthritis, Newcastle disease, avian paramyxovirus infection, highly pathogenic avian influenza and West Nile fever were mentioned, respectively. Especially, since the WNV has been reported from the Far East, the virus will invade to Japan in nearly future. According to the computer simulation of the risk assessment of extinction due to an outbreak of West Nile disease among free ranging Red-crowned Crane (*Grus japonensis*) in Hokkaido Island, Japan, with usage of the software "Vortex", the population will die out within 40 years. Hence, the

VecTest has been performed on oral and tissue swabs from all wild birds carried to the Wild Animal Medical Center in the Teaching Animal Hospital of the School of Veterinary Medicine, Rakuno Gakuen University. In bacterial diseases epidemiology, systematic surveillance and/or individual case reports on *Esherichia coli, Actinomyces* spp., *Plesiomonas shigelloides*, *Staphylococcus epidermidis Salmonella enterica* serotype Typhimurium DT40, *Vibrio cholera, V. parahaemolyticus, Campylobacter jejuni, Chlamydophila psittaci, Mycobacterium genavense* etc were given. Above all, mass mortality of sparrows (*Passer montanus*) due to an epidemic caused by *S. enterica* has been just started from 2005 to now. And, the mortality has seemed to be often associated with a chemical deicer poisoning and/or staphylococcal infection, respectively. Although diseases due to biotoxins are defined as a non-infectious disease, some of the diseases have an intimate relationship with presence of certain responsible bacteria, e. g., cyanobacteria, such as the genera *Microcystis*, *Anabaena, Aphanizomenon, Nodularia*, and *Oscillatoria*. And, some suspicious cases of such problems were recorded even in Japan. Several fungal genera, in particular *Candida, Cryptococcus, Aspergillus, Microsporum, Trichophyton, Fusarium, Ochroconis* and *Absidia* have been reported as causing diseases as well, and some suspicious cases in free-ranging birds without a positive identification of each responsible fungal species have been given. Furthermore, there are several genera of avian flagellates and amoeba (Phylum Sarcomastigophora), such as the genera *Trypanosoma, Hexamita, Histomonas, Parahistomonas, Monocercomonas, Trichomonas, Tetratrichomonas, Chilomastix, Entamoeba*, *Endolimax* etc. In particular, several cases due to *Trichomonas*, *Tetratrichomonas* and/or *Hexamita* infection have been found in wild birds, but there have been no reported mass mortality due to the protozoan agents in Japan. And, *Trypanosoma* spp. have, however, been found in wild and/or captive passerines and owls, but they seem to have been non-pathogenic, at least now. On the other hand, some of avian coccidian and haematozoan genera (Phylum Apicomplexa), such as *Eimeria, Isospora, Tyzzeria, Wenyonella, Caryospora, Cryptosporidium, Sarcocystis, Toxoplasma, Atoxoplasma, Hepatozoon, Haemoproteus, Leucocytozoon, Plasmodium* and *Babesia*, may be responsible for severe disease. First of all, several cases including observatrion of *Isospora* shaped oocysts of *Atoxoplasma* sp. from fecal examinations and/or histopathological observation from postmortem examinations from not only captive Bali Myna (*Leucopsar rothschildi*), but also free-ranging passerines, e.g., *Passer montanus, Chloris sinica* and *Spodiopsar cineraceus.* There have been many case reports on intestinal coccidiosis including *Cryptosporidium meleagridis* infection from common avian species, it is prominent that *Eimeria ueki* was found from endangered subspecies, *Lagopus muta japonica*, on high mountain area. In 2002, fatal cases of the Red-crowned cranes due to disseminated visceral coccidiosis, caused by *Eimeria reichenowi* and *E. gruis*, were found. Control of the visceral coccidian species is more difficult than the control of intestinal coccidian species. Adding to this, the authors could observe renal coccidiosis concomitant infection with *Corynebacterium* sp. and *Renicola* sp. from mass mortality of *Puffinus tenuirostris. Sarcocystis*, most commonly found among waterfowl in the United States, is nonfatal, but severe infections can cause loss of muscle tissue and result in lameness, weakness etc. Recently, there have been several records of the protozoans among wild birds in Japan as well. Within the areas of both public health and captive animal health, toxoplasmosis among poultry is also very important. Serological evidence of the protozoan infection has been found so far among almost one third individuals of free-ranging anatids in Obihiro, Hokkaido Island. Although the agents for avian malaria including the genera

Leucocytozoon, *Haemoproteus* and *Plasmodium* have been commonly found among various avian species including endangered ones, e.g., *Ketupa blakistoni, Haliaeetus pelagicus, Lagopus muta japonica* etc and exotic one, e.g., *Leiothrix lutea*, in Japan. Adding to the protozoans, at least in Japan, hosts of the genus *Babesia* have been restricted to mammals, but the piroplasma was obtained from *Spilornis cheela* in Yaeyama Archipelago, southern part of Japan. Certain free living protozoans such as diatoms (Phylum Bacillariophyta: genera *Nitzschia* and *Pseudonitzschia*) and dinoflagellates (Phylum Dinoflagellata: genera *Alexandrium* and *Gymnodinium*) produce neurotoxins such as domoic acid, saxitoxin, and brevetoxin, that can affect sea birds. Most reports of mortality in birds have been die-offs occurring in conjunction with protozoan blooms in the United States, but no such mortalities have been reported yet in Japan. Nevertheless, many protozoan blooms occur around Japan making it essential that the authors pay attention to such events because of their potential for causing mortality. Including the diseases, the disease outbreak or mass mortality prevention will need to be based not only on an understanding of avian ecology but also of agents (pathogens) epidemiology. And, in order for medical researchers to be able to prepare for future epidemic outbreaks amongst birds, they must perform epidemiological surveys of the probable agents mentioned above in wild bird populations both in Japan and its surrounding areas; international and inter-agency cooperation is, therefore, essential in disease prevention. Needless to say, very careful attention should be paid to the possibility of infection by zoonotic agents during monitoring surveys.

Chapter 6 - Brazil has one of the richest avifauna in the world. There are 1,832 species distributed in 31 orders and 98 families. The Amazon, Atlantic Forest and Cerrado biomes have the richest diversity of birds in the country, but the most endemic bird species are found in the Atlantic Forest. Several species of migratory birds are also found in this country, originating from North and Central America and some countries in South America. Therefore, data from studies focusing parasitism by ticks and the pathogens they transmit are of great importance to animal and human welfare. To date, 6,201 ticks (4,270 larvae, 1,784 nymphs and 147 adults) have been collected. Passeriformes was the most parasitized order (5,964 ticks). The Thamnophilidae shows the highest numbers of parasitized species 25 (14%) out of 174 occurring in Brazil, followed by Tyrannidae (22/13.16%), Furnariidae (18/137. 17%) and Thraupidae (17/99. 17%). A total of 21 species of ticks belonging to four genera: *Amblyomma*, *Rhipicephalus*, *Ixodes* and *Haemaphysalis* have been identified. The most abundant tick species are *Amblyomma longirostre* and *Amblyomma nodosum*. To date only bacteria from the genus *Rickettsia* are isolated from ticks associated with wild birds: from *Amblyomma longirostre, Rickettsia amblyommi* and *Rickettsia parkeri;* from *Amblyomma nodosum, R. parkeri;* from *Amblyomma geayi, R. amblyommi;* from *Amblyomma parkeri, R. parkeri,* from *Amblyomma calcaratum, R. parkeri* and from *Amblyomma sp., R. parkeri.*

Chapter 7 - Most parasitic animals including helminths, e. g., trematodes, cestodes, nematodes, acanthocephalans, leeches etc., and arthropods, e. g., ticks, mites and tongue worms, are component parts of the ecosystems in which their wild avian hosts occur, they have co-evolved, and the infectious agents do not necessarily cause disease. Sometimes, however, some individuals become sick and in extreme cases die, in various parts of the world because of an outbreak of an epidemic disease. And, considerable environmental change has occurred in Japan, especially, most northern part of Hokkaido Island, over the last 150 years as well. Furthermore, the island is one of important transit sites of the country for a wide range of migratory avian species, because that is situated on the East Asian Flyway a migratory route. Hence, the mortality or subclinical diseases due to the parasitic animals mentioned above have not so far significantly impacted populations of birds only in Japan. Because an understanding of the parasitic avian disease organisms that are present in Japan, their ecology, transmission, and symptoms is an essential conservation tool, the authors have overviewed this field by the end of the 20 century. However, according to the decade passed by, several issues have to be added to the review paper. So, Yoshino and Asakawa (2013) presented the nematode diseases and/or agents recorded from anatid species on the island. Hence, this chapter provides a recent overview of the parasitic diseases and/or their responsible agents recorded on non-anatid avian species as the succession, and briefly potential strategies for epidemic risk reduction are given based on the data from Wild Animal Medical Center, Rakuno Gakuen University. By the present, parasitic helminths including total 175 trematode, 148 cestode, 70 nematode, and 20 acanthocephalan species have been recorded. Among trematodes, *Leucochloridium* sp., *Renicola* sp., *Balfouria monogama* etc were found. Several cases of nematodiasis including fatal cases were reported, for example, eustrongylidosis was confirmed as one of factors contributing to the mortality of such piscivorous birds in Japan, too. The spirurid belonging to the genera *Thelazia*, *Acuaria, Cosmocephalus, Dispharynx, Synhimantus, Tetrameres* etc. have been found among various avian species including endangered ones by the author's recent studies. As other nematodes, e. g., genera *Heterakis*, *Porrocaecum*, *Strongyloides, Hamatospiculum, Desmidocercella* etc were reported from not only intestines, but also air sac. Several acanthocephalan, bloodsucking leeches, e. g., *Theromyzon*, external/internal parasitic arthropods could affect many free-ranging birds. The parasitic diseases prevention will need to be based not only on an understanding of avian ecology but also of parasites. And, in order for medical, conservation ecological and/or veterinary medical researchers to be able to prepare for future epidemic issues amongst birds, they must perform epidemiological surveys on some highly pathogenic parasites mentioned above in wild bird populations in Japan including Hokkaido Island. In short, an international cooperation is essential in parasitic disease prevention. And, furthermore, proper educational activities for a citizen of the area are needed for the trial as well. For example, the authors know that parasites from a wild bird may be a popular exhibition to general citizen, because of showing indirectly an ecological aspect of the bird species. That is very important for the potential strategies against the parasitic disease control because of a holder of the casting vote for the avian future is eventually such people.

Chapter 8 - In the present study the authors determine in to x35 areas of Iberian Peninsula along a gradient of urbanization the following factors: After the age of the park (AG), altitude (AL), number of buildings in a circle of 500 m (BL), cover of grass (CG), shrubs (CS) and trees (CT), average noise (DB), distance to nearest city (DC), use of herbicides (HB), inhabitants of nearest city (NC), number of persons in the park (NP), density of old trees (OT), park area (PA), species richness of shrubs (RS), trees (RT) and woody plants (RW) and number of water areas (WA). The authors check the relationships of these environmental factors with the total bird abundance (N), the alpha-fisher diversity (α), the bird species richness (S), the Berger-Parker dominance (d), the Margalef diversity (Mg), the Shannon diversity (H), the Shannon entropy (e^h), the Simpson diversity (D) and the Simpson dominance (1-D) using stepwise multiple linear regression. To validate these statistical models, the authors determine the coefficients of determination (R^2), the residual normality by Shapiro-Wilk test (W) and the residual homoscedasticity by Breusch-Pagan test (BP). In all the regressions the authors obtained exhigh coefficients of determination, residual normality (except in Simpson dominance) and residual homoscedasticity. The main causal variables that effected ecological parameters of community were altitude, distance to nearest city, noise level and shrub cover. The age of the park was irrelevant because in urban areas the environment is continuously modified by gardening and, consequently, the succession processes are not produced. The author's study shows that consequently, urban parks could be considered potential reserves for bird conservation if the authors manage these four factors in an optimal way.

Chapter 9 - The Hooded Vulture, the smallest and commonest vulture in West Africa, is also possibly the most adaptive African vulture, evidenced by its tolerance of human proximity and high urban presence. However, very few studies assess its role as a dominant urban and rural scavenger in West African contexts outside the range of the larger vultures, rather focusing on its relationship with the larger vultures in the Eastern and Southern African plains. This chapter examines the relationships between Hooded Vultures, competing avian scavengers (black kites and pied crows) and rural/urban landscape configuration in the little researched southern half of Ghana, West Africa. Similar to crows and kites, hooded vulture numbers were positively correlated with settlement populations, and these birds had a stronger presence in urban than rural contexts. Hooded vultures had higher correlations with organic waste than crows or kites, and were the commonest avian scavengers in urban areas with high organic waste (abattoirs, markets), but less common than pied crows in residential and business districts with low organic waste. Vultures, kites and crows were commoner in urban areas near dense forest (including dense, urban green spaces), than in urban areas near lighter vegetation. Vultures nesting was correlated with dense rural forests, and to a lesser extent with dense, urban green spaces. Vultures were active in the late evening, early dawn and night in urban areas. In rural areas, vulture numbers were less correlated with vegetation density than in urban areas, and town proximity was positively correlated with vulture numbers. Conflicts between vultures, kites, crows and feral dogs were common, and no species was significantly dominant. The hooded vulture therefore has some specialist tendencies, despite its common portrayal as a successful generalist.

In: Birds
Editors: Lucas Ruiz and Franco Iglesias

ISBN: 978-1-62948-104-3

Chapter 1

Sex Determination in Birds

***A. F. Smirnov** and A. V. Trukhina*†**
Department of Genetics and Biotechnology, St.-Petersburg State University,
St. Petersburg, Russia

Abstract

How gender is formed in birds? What caused dimorphism and formation of biologically important feature? What are the internal physiological mechanisms and the role of the genome in the process? To all these questions are answered offered to the reader head chapter. Paper is devoted to sex determination in birds. It will be reviewed data on the genome of birds, information on the sex chromosomes, heterogametic, sex of birds of the different karyotype. Specificity of Z and W chromosomes, sex chromosomes in Ratinae will be considered. The ideas will be presented about the role of hormones in sex determination and differentiation. The data on the possibility of "inversion" of sex in birds in the inhibition of aromatase, estrogen exposure and gonadectomy will be introduced. In addition, review key sex determining genes, chromosomal determinants, specific dosage compensation will be discussed. Sex determining genes will be examined in birds, including Dmrt1, AMH, FOXL2 and W chromosomal determinants. The specificity of birds dose compensation will be described. The data will be submitted about MHM region of Z chromosome. Features of sex determination in males and females will be discussed. Bird cell-autonomous sexual identity (CASI) will be considered.

Introduction

In therian mammals (marsupials and 'placentals') sex is set at fertilization by the inheritance of the sex chromosomes. Embryos inheriting two X chromosomes develop as females whereas those inheriting an X and a Y chromosome develop as males. The *SRY* gene

* E-mail: afsmirnov@bio.pu.ru.
† E-mail: trukhina_ant@mail.ru.

located on the Y chromosome represents the master sex determinant in higher mammals. However birds lack *SRY* and have a non-homologous set of sex chromosomes designated ZZ male and ZW female. This article will summarise the current knowledge of avian sex determination. Avian sex chromosomes will be described and the possible mechanisms of sex determination will be considered.

Sex Chromosomes of Birds, Sex Chromosomal Abnormalities

Similar to all animals in birds there are chromosomal differences between the sexes. There are two Z chromosomes are in males, while females have got Z and W chromosomes, therefore heterogametic sex is female. This type of definition of sex exists in many species of reptiles, *Lepidoptera* and birds. Known case of the presence of two systems of sex determination is XX / XY and ZZ / ZW within a species is in the frog *Rana raposa*. It is thought that these two systems either have a common evolutionary origin or XY system of mammals arose directly from the ZW system of reptiles [1].

Lets look in greater detail how sex is determined in birds. Z chromosome in these species is not subject of inactivation and it has many genes. About 20% of anonymous DNA probes were mapped here. W chromosome is similar to the mammalian Y chromosome that is almost entirely heterochromatic, late replicated and paired with a small part Z chromosome during meiosis. The avian Z chromosome is large and gene-rich (similar to the mammalian X chromosome) while the W is typically small and largely heterochromatic (similar to the mammalian Y). The chicken Z chromosome is about 82Mb and is highly conserved among avians [2, 3]. The Z harbors over 1,000 genes [4]. 60% of the W chromosome DNA sequence consist of repeats such as Xho1 (centromeric region) and EcoR1 families (large region of the long arm and the middle of the short arm). There is also Ssp1 family containing a repeating unit of 0.5kb, including polypurin / polypyrimidine sequence (GGAGA repeat) at the 3'end. Interestingly, this family forms a more diffuse chromatin in the nucleus compared with EcoR1 and Xho1 repeats and it is believed that it is a kind of barrier distribution of heterochromatin in the terminal euchromatic region. The sizes of these families are about 6 Mb (Ssp1), 11 Mb (EcoR1) and 21 Mb (Xho1) respectively. There is about 6 Mb of uncharacterized repetitive DNA in this chromosome. W chromosome has got approximately 44Mb of repeats and about 10Mb of unique DNA, which may be consists 20 active genes (Figure 1), [5]. RNA-seq analysis allowed the compilation of twenty-six W-linked genes. Most of the expressed W genes have open reading frames that are either almost or completely homologous to their Z-linked gametologues. The two exceptions are *HINTW*, which is divergent from its Z gametologue, and *FAF*, which lacks a Z gametologue [6]. Only 4% of the chicken W chromosome DNA sequences have been mapped [7]. This amounts to around 1.2Mb of assembled sequence on the W in the most recent release of the chicken genome [8]. Importantly, the longest telomere beams of Class 3 (2.8 Mb) is localized to the long arm of the W chromosome. This phenomenon is associated with recombination hot spots, whose role is to be a specific buffer for gene-rich regions [9].

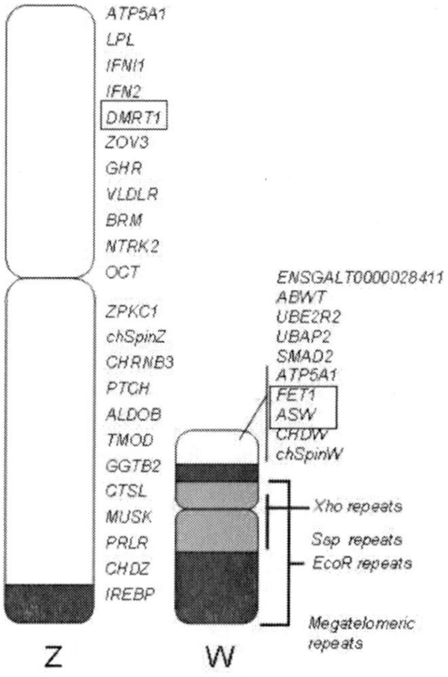

Figure 1. Z and W chromosomes of *Gallus gallus domesticus*. Euchromatin is showed white color, repetitive sequences (heterochromatin) is showed black and gray color. Also it is showed candidate gene for male (*DMRT1*) and prospective sex determined genes for female (*FET1* and *ASW*).

In *Carinatae (Neognathae)* Z and W chromosomes are clearly distinguishable. In hawk family *Accipitridae* W chromosome has interspecific variability. In division *Ratitae* (ostrich, emu) sex chromosomes are only slightly differentiated. It is thought that sex chromosomes in birds and mammals are not homologous and evolved from different pairs of autosomes. So Z chromosome of many birds contains homologous regions of chromosome 9 (the most of the 24 namely 17 mapped on chromosome 9), 5, 8, 18, and there is only one site of homology of X chromosome containing the gene *OTC* (Ornityhine transcarbamylase) (**Figure 2**), [2].

It is believed that ancestral Z chromosome appeared after the removal of bony fish and tetrapods (terrestrial vertebrates) because there is no conservative synteny of Z chromosomal genes in zebrafish and puffer fish. Comparative mapping also demonstrated a connection between mammal X chromosome short arm and chicken chromosome 4. A small fragment of the X chromosome has also homology to chicken chromosome 12 [2].

It is assumed that the W chromosome is formed from some proto-Z chromosome by series of inactivations, inversions and degradations, accumulation Z and W-specific repeats for this particular presentation at the Z and W chromosomes of two genes *Ratitae ACO1*, *ZOV3* and pseudogene similar sequence EE06. Conservative karyotype was described in turtles and birds. Six large autosomes are almost equivalent. Affinity with snakes is less presented. Long arm of chromosome 6 of turtle and short arm of chromosome 2 of one snake have a synteny with Z chromosome of birds. Turtles have got XX / XY and ZZ / ZW mechanisms of sex determination. Snakes have only ZZ / ZW sex-determination system. And their Z chromosome has a different origin [10, 11, 12].

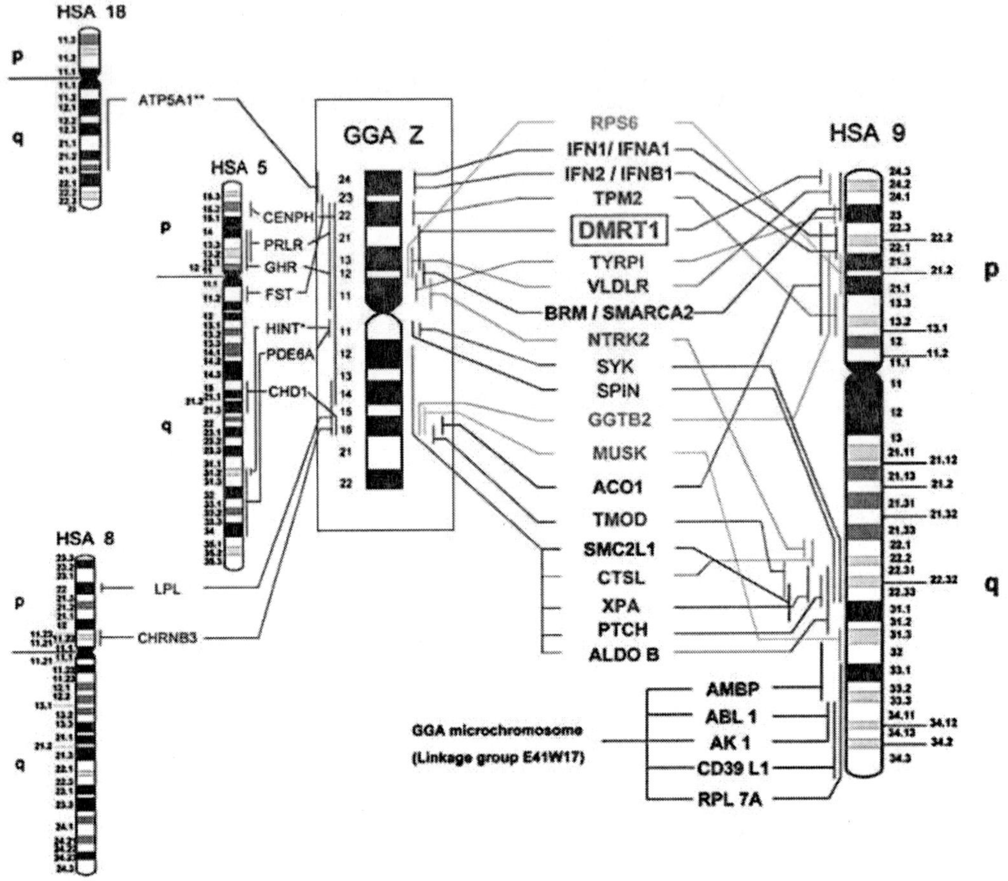

Figure 2. Genes mapped to the chicken Z chromosome (GGAZ) and their localization on human chromosomes (HSA). Vertical bars indicate the physical location of genes on the Z chromosome.

The evolution of heterogametic chromosomes Z and W is similar to that of X and Y in mammals. Progressive restriction of recombination is an integral feature of chromosomal evolution in female heterogametic. There are three evolutionary strata on the Z chromosome created 132-150 (stratum 1), 71-99 (stratum 2) and 47-57 (stratum 3) million years ago (MYA). Most neognath birds have got very short PAR (psevdoautosomal region) in Z chromosome and it is rather long at paleognath [13, 14]. All birds have homologous sex chromosomes that formed about 120 million years ago, similar in age to the mammalian sex chromosomes (about 165 millions years ago). However, whereas all mammals and most bird lineages have highly differentiated sex chromosomes, in some groups of birds such as ratites the sex chromosomes remain homomorphic. It is interesting that the emu (*Dromaius novaehollandiae*) sex chromosomes have become masculinized, even in the absence of ZW differentiation [15].

SRY gene in birds is not identified. It is believed that sex of these species can be controlled. In the first the W sex chromosome carries a dominant - acting female determinant. In the second sex could be determined by dosage of one or more Z-linked genes, i.e. two copies in males and one in females. In the third it is possible both of these mechanisms exist [6]. That is, the sex determination may be follow up by type of fruit fly (balance theory) or by type of mammals. Individuals with crucial genotypes 2A: ZO and 2A: ZZW have not been practically described. Animals with karyotype 3A: ZZZ are always males but sterile. The available experimental data indicate, perhaps, in favor of the balance. 3A: ZZW-intersex individuals have the right gonad, testes, and the left one ovo-testis (further developed in the testes). Thus should be the ovary, if W has got ovary determining gene. Individuals 3A: ZWW perished on day 16, but they have formed ovary.

Gynandromorphy (right-testes, left ovary) are described in parrots, respectively, with ZZ karyotype and Z0. The 2A:ZZW individual of a reed warblers (*Acrocephalus arundinaceus*) was recently described, behaving as a female in a natural population. Genetically, it was proved that it has two Z chromosomes (two different alleles *G61* and *Aar1*). Moreover only one particular maternal allele is transmitted male offspring (n=12) and is not given to females (n=5). It is assumed that the female was trisomal (2A:ZW) in somatic tissues and diploid in sex [15, 16, 17]. In 2012 fertile triploid ZZW female of plovers (*Chavadvius alexandrines*) was described. This points to the special role of W chromosomes in sex determination in birds [18, 19]. **Figure 3** presents the data about sex of aneuploids birds in terms of the hypothesis of a W dominant gene on sex determining W chromosome and dose sensitive of Z chromosome.

Taken together these studies imply that the presence of the W chromosome is associated with at least some degree of ovarian tissue development; however it appears that this can ultimately be over-ridden by two Z chromosomes.

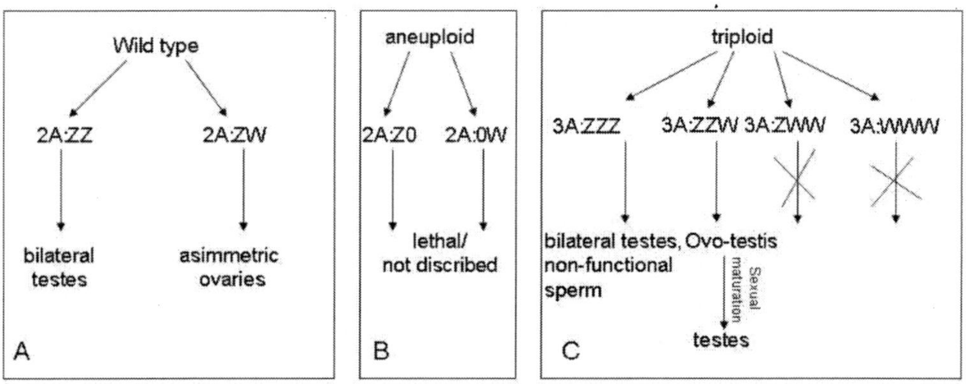

Figure 3. The role of sex chromosome in chicken aneuploids. The exact mechanism through which sex is determined by the chicken sex chromosomes is unknown. In normal birds females are heterogametic (ZW) while males are homogametic (ZZ). A.) Studying aneuploids might help shed light on meaning the existence of a Z dosage mechanism or dominant W mechanism of sex determination that have been suggested for birds. Sex chromosome aneuploids have not been described and are thought to be lethal. B) Chickens with triploidy (including the sex chromosomes) have been extensively described. C) In triploid male of birds (ZZZ) the bilateral testis are developed normally although sperm was abnormal. ZZW triploid birds were initially female-like and had ovotestis, but the ovarian tissue was regressed with time and these birds presented with more male-like features after sexual maturity.

Development of Gonads in Birds

In birds as in mammals development of gonads occurs on the surface of genital mesonephros. Next step is the primary sex determination: the formation of the seminal tubes, ovaries, steroidogenic cells. It is thought that birds, unlike mammals, have no undifferentiated bisexual but bipotential primary gonad and sex determination occurs earlier than in mammals (**Figure 4**), [20].

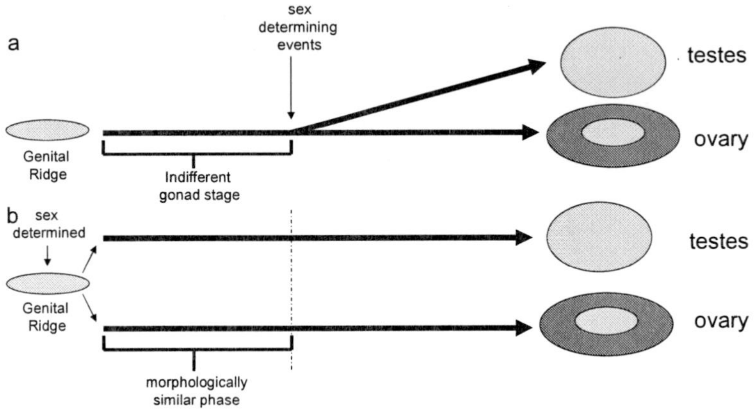

Figure 4. A possible model of development of the gonads in birds (b) compared with mammals (a). It draws attention to the lack of birds bisexual (indifferent) stage.

Gonad development in chickens was studied more fully. Full-embryogenesis is 21 days. Gonad appears at 3.5 day (22 steps) as a thickening on the surface of the mesonephros. It consists of the epithelial layer of somatic and germ cells, and medullary cordate layer mixed with the cells of the mesenchyme. The first sexual differentiation is observed on 6,5 day (30 step). At this stage the main sex-determining genes start to operate. Sertoli cells begin to differentiate and begin to appear vas way of males and outer cortical thickening gonads in females. In ZZ individuals within medulla appear Sertoli cells and bilateral testes with seminiferous tubules and with germ cells within them. It is observed reduction of the cortex. In ZW embryos left gonad develops into an ovary. There is fragmentation of medullar hypertrophy of germinal epithelium. Right gonad initially differentiated as ovary through fragmentation and thickening medullary cortex (**Figure 5**).

So gonads form at around embryonic day 3 and are initially bipotential. Prior to E6 in the male, the Z-linked transcription factor *DMRT1* becomes highly expressed in the male gonads, initiating testis development which begins morphologically differentiate at around E6. This may be through regulation of genes such as *SOX9* and through repression of female pathways. It leads to the development of bilateral testis, characterized by seminiferous cord structures. Differentiation is mostly complete by E9. In the female *Foxl2* becomes expressed at around E5, and this is thought to regulate the expression of the aromatase gene which is important for estrogen synthesis. A second parallel pathway, the *RSPO1/Wnt4* pathways, is also upregulated in female gonads, and may play a role in ovarian development which includes fragmentation of the medulla to produce lacunae and proliferation of the cortex cells in the left gonad. Nonproliferation and subsequent regression of the right gonad leads to asymmetric gonads in the female (**Figure 6**), [5, 6, 21].

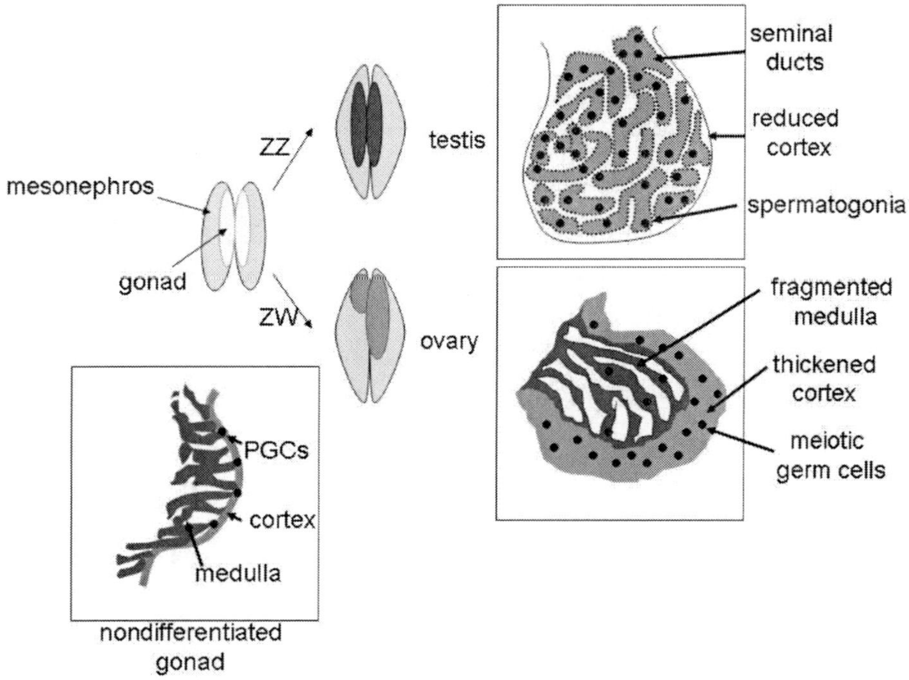

Figure 5. The gonadal sex differentiation in chicken embryo. Primordial germ cells (PGCs) are visible on 3.5 day on the ventromedial surface of the mesonephros kidneys. Gonad is undifferentiated. It consists of an outer cortex and underlaying medullary layer. Mesenchymal cells distributed between the cord and medulla. PGCs are located mainly in the cortex. On 5.5 - 6 day (stage 28-30) in ZZ embryos appear bilateral testes. In ZW embryos left gonad gives ovaries and the right one regresses. Medullary cord initiated appearance seminiferous tubules. It is fragmented with the emergence of gaps in the developing ovary. Cortex is reduced in testes, but develops and proliferates in the ovaries.

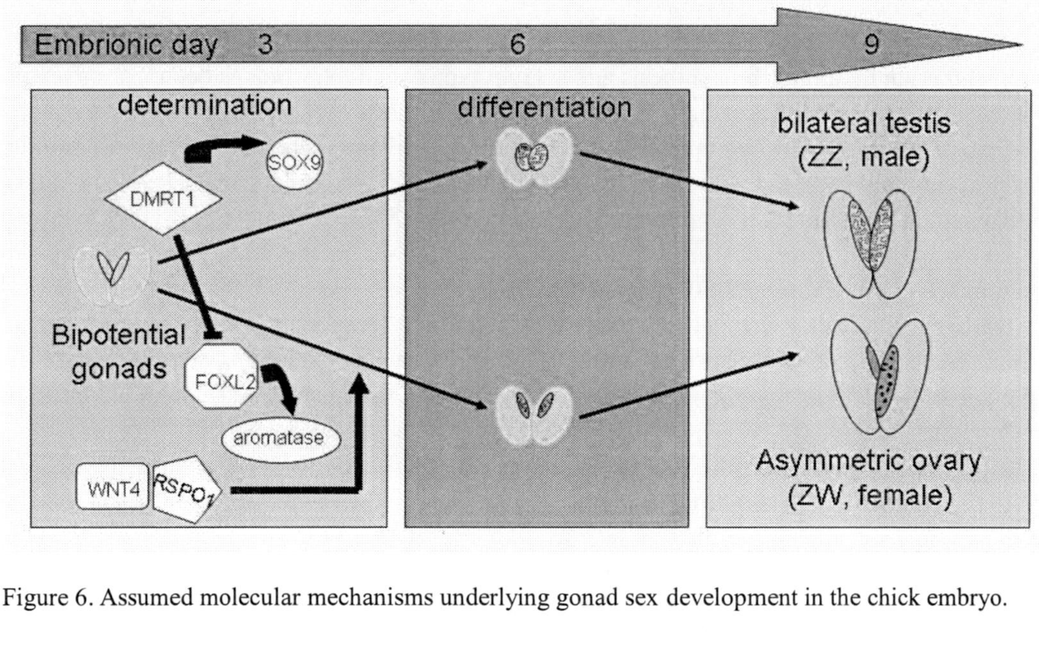

Figure 6. Assumed molecular mechanisms underlying gonad sex development in the chick embryo.

Hormones and Sexual Differentiation in Birds

Estrogens is both necessary and sufficient to drive ovarian development in many nonmammalian vertebrates (**Table 1).** However, the role of estrogen in the mammalian gonad is less clear. Birds and marsupials are unique because their sex is determined by the composition of the sex chromosomes, although the embryo remains sensitive to the effects of estrogens. It is especially so for the chicken (*Gallus gallus)* and tammar wallaby (*Macropus eugenii*), the most characterized models for avian and marsupial species, respectively. Although birds diverged from reptiles approximately 100 million years ago, the ovary-determining action of estrogens remains present in birds with the evolution of a genetic mechanism for sex determination [22, 23, 24].

Table 1. The role of estrogens in sex determination of vertebrates and phylogenetic distance [24]

Group of vertebrata	Distance from mammals (million years ago, MYA)	Influence of estrogens on sex determination
Nonmammalian vertebrata		
Pisces	450-530	+
Amphibia	300	+
Reptilia	290	+
Aves	216-199	+
Theria		
Monotremata	180	+
Marsupialia	160	+
Eutheria (Mammalia)	0	—

In eutherian mammals biosynthesis and secretion of gonadal hormones occurs after sex or gonadal differentiation. According to Pask A.J. [23] in the presence of estrogens key male differentiation genes fail to be upregulated in the XY gonad and instead key ovary-promoting genes are upregulated leading to ovarian development. For birds of gonad development is sensitive to the manipulation of hormones. Two enzymes - P450 aromatase 17β-HSD required for the synthesis of estrogens only expressed in ZW gonads in the stage of morphological resolution (6-6.5 day, stage 29-30) (**Figure 7).**

Estrogen synthesis is critical for sexual differentiation. Its receptor (ER-α) is synthesized in gonadal cortex of both sexes. A proper estrogen synthesizing enzyme aromatase is expressed only in female gonads (medullar cord). The first time it was detected at 6.5 days of development (stage 30). When ZW embryos process with aromatase inhibitor embryos develop bilateral testes. Another enzyme involved in the biosynthesis of estrogens is 17β-HSD (17β-hydrosteroid dehydrogenase). It is also expressed in embryos medulla of female embryos, but not males. It also points to the importance of estrogen in gonadal sex differentiation in birds. Estradiol is synthesized in the gonads of females and medulla mediates sex determination through the stimulation of the cortex with ER-α [10].

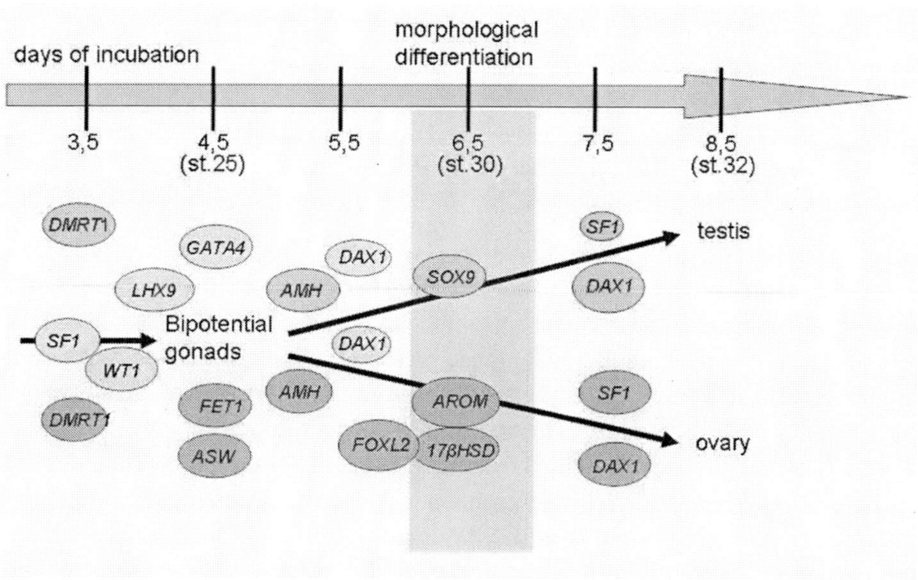

Figure 7. Expression of key genes during the development of gonadal sex in chick embryos. Morphologically undifferentiated gonad to 6,5 days of embryonic development (stage 29-30). The rings represent specific expression of genes showing sex specificity. Their size corresponds to the level of expression.

That is the differentiation of gonads is sensitive to the effects of estrogens and androgens in fish and amphibians. There are the effects of estrogen in reptiles, birds and marsupials. In noneutherian vertebrate gonads are formed outside the mother's body, unlike mammals. In mammals aromatase is expressed later in embryonic development and gonadal sex is formed. The differentiation can occur in the absence of steroidogenesis regardless of sex hormones [22-24].

It should be noted that the AMH (MIS antimullerian hormone) inducing regression of Mullerian ducts in male mammals is synthesized in the Sertoli cells. It is expressed after the development of the testes and key sex-determining genes expression such as *SOX9*. It is not required for the development of testes, based on experiments on mutagenesis. The expression of chicken *AMH* begins before sex differentiation (**Figure 7**). Adding AMH can induce sex reversal in chicken females. AMH may play a more important role in sex determination (male-determining) in birds compared with mammals possibly by interfering with aromatase [21].

In chick sexing occurs somewhere at 5.5 days of development. For all birds characteristic asymmetry gonads can be observed: females left ovary is fully developed, the right gonad regresses after pipping, and males the left testis greater than the right one. On 7 days of incubation in the left and right gonad expressed all five genes required for the synthesis of estradiol (estradiol-17 β) from cholesterol. But the enzymes that catalyze the last two steps of synthesis of estrogens (17β-HSD and P450-aromatase) are expressed at very low levels in the right and left male gonads. Estradiol is produced in the left and regressing right gonad of females but not produced at a significant level in the gonads in males. Interestingly, the mRNA for the estrogen receptor (ER) is observed at stages 28 and 29 (6 days of development) in the left gonad of males and females. From stage 28 to 38 its level is limited in the left gonad in males. Males and females begin to differ to 7.5 day incubation. Levels of testosterone in males and estradiol in females are correspondingly higher [21, 25, 26].

Sex-Determining Genes in Birds

Candidate for the role of sex-determination above all we must mention the genes present in the Z and W chromosomes or both. This is *CHD* (Chromodomain-Helicase-DNA-binding protein) gene. Birds have Z and W chromosome homologues (*CHDZ* and *CHDW*). These genes are conserved and homologous mouse helicase which regulates transcription by modifying chromatin structure. The expression pattern of this gene during the development of the gonads has not been investigated. Interestingly the heteromorphic sex chromosomes and *CHD-W* are absent in the running emu, ostrich, nandus. The gene is not universal for sex determination. In *Ratitae* versus *Neognathae* the sex is defined differently [27].

Another gene *ASW* (Avian Sex-specific, W-linked) / *PKCI-W* localized on chromosome W (again, except for the running). It is one of the main candidates for sex determining. There are 40 copies of the gene in the W chromosome and one copy *PKCI-Z* in Z chromosomes. At the cDNA level the similarity between the Z and W chromosome homologues PKCI is about 65%. The structural differences between these genes are the absence of motive *PKCI-W* HIT (histidine triad) which is essential to the function of the protein kinase. Enzymes with HIT motif function as homodimers and *WPKC1* functions as a regulator work through the formation of a heterodimer. Heterodimer formed *PKC1-Z* and *PKC1-W* inhibits the activity of female *PKC1-Z*. In individuals ZZW more HINT homodimer is formed and partly overcome feminizing effect of *ASW*. Hint enzyme activity is important for the appearance of ZZ males and its inhibition appears to ZW female. Both of these genes are expressed in the early development stage of the females to 30 and is represented on the stage 26-27 on the left and right gonad, mesonephros globules, early neural tissue. Gene HINTW titrates HINTZ, blocking the function of histidine triad in HINTZ, leading to loss of AMP binding. The same *HINTZ* gene is potential regulator of transcription of certain genes such as transcription factor *MITF*. There are some counter-arguments to the *HINT* role as sex determining gene: 1) its activity is shown on day 2 of differentiation but against sexual differentiation on day 6; 2) there is evidence for the fact that the HINTW activity is not reduced by sex inversion with aromatase inhibitors; 3) *HINTW* mRNA is detected in all extraembryonic tissues; 4) *HINTW* gene is absent in Z chromosome running birds separated from *Neognathae* about 80 million years ago [26, 27]. The use of retroviral vectors for *HINTW* gene transfer experiments showed that despite of expression in gonads ZZ embryos develop into testes. This result argues against a decisive role in *HINTW* sex determination in birds [28, 29, 30].

Recently identified gene *FET1* (Female expressed transcript 1) is localized in the euchromatic part of the W chromosome of chicken and it has not got Z chromosome homologue. It is expressed exclusively in the female genitalia (most strongly on days 4.5-6.5 (stage 25-30)), mainly in left gonad. There is no homology with any known gene. The gene encodes a transmembrane protein of 494 amino acid residues. It is assumed that this gene influences on subsequent expression of aromatase. It can be associated with factor regulating dosage compensation in chickens (F factor or *ZUF* - Z upregulating factor) [31].

Birds do not have *SRY* as a main sex determining gene of mammals. However there are other mammalian sex-determining genes like *AMH*, *SOX9*, *SOX3*, *WT1*, *SF1*, *DAX1*.

SOX9 gene begins to be expressed in male gonads at 6,5 day (30 stage). In birds, however, the *AMH* expression precedes SOX9 activity. At a time when the gonads begin to differentiate gene is expressed exclusively in the testes [32, 25].

DMRT1 gene (DM-related transcription factor 1) is localized in Z chromosome in birds. It is expressed in the developing embryo in the urogenital system as well as in mice. Ortholog was identified in short arm chromosome 9 humans and it is responsible for the appearance of individuals with chromosomal inversion of gender. It is related *Drosophila* gene *dsx* (doublesex) and *mab3* gene (male abnomal 3) of nematode (*Caenorhabditis elegans*). The proteins encoded by these genes contain DNA binding motif (DM-domain). In mice *Dmrtl* gene is expressed exclusively in gonadal tubercle in the XX and XY embryos and then with the development in the seminiferous tubules, testes, Sertoli cells, germ cells. Violating of its activity gives rise to XY mice with defective testes. In chick expression associated with medulla intense in the gonads in males and appears at 4.5 - 5.5 day (stages 25-28) of development and even at 3.5 days of development (stage 22). This gene is expressed in gonadal strip of birds (for *SRY* is only typical) during its formation and generation Wolffian ducts. Its activity is more in male embryos than female. Gene is located at the site of Z chromosome cytogenetically coinciding with MHM region and hence the assumption of its role in dosage compensation and sex determination in birds. *DMRT* gene is expressed at day 3 of development and the differentiation of sex to the histological level begins on day 6 of embryonic development. RNA interference mediated suppression of *DMRT1* lead to feminization of the testis at both structural and molecular level. It should be noted that sexual dimorphism in gene expression of the birds occurs before histological differentiation of the gonads. *DMRT1* is strongly expressed in males at 3.5 day of development (22 stage) for morphologically undifferentiated gonads. *AMH* is also working on stages 25-28 before differentiation. At the molecular level differentiation of bird gonads occurs earlier than morphological. [33, 34, 35].

DMRT1 gene was recently cloned in emu and localized in short arm Z chromosome but it is not in W. The sequence of this determinant has got 88% homologous to the corresponding gene of chicken and 65% of human. 270bp sequence in intron 3 of *DMRT1* gene emu has 90% homologous to the corresponding human intron. This high conservatism suggests some important role in controlling the expression of *DMRT1*, for example, control of methylation in females. Two copies of the gene are likely required for the appearance of testes in ZZ males [36].

Aside from *DMRT1* and *SOX9,* expression of several other important mammalian testis factors have been investigated in the chicken. *SOX3* gene worked until *SOX9* expression and somehow regulated in both sexes until the formation of gonads. *SF1* gene belongs to the large family of nuclear hormonal receptors. Its ligand is not yet identified. It is expressed in the undifferentiated gonad of both sexes in mice and it regulates the development of the undifferentiated gonads and male specific differentiation. In birds this gene is expressed in differentiating ovary much stronger than the testes. Its expression starts under 6.5 day of development (stage 30) and runs up to 8.5 days of development (stage 35). Its activity is associated presumably with high steroidogenic biosynthesis in females and the possible activation of aromatase. In any case *SF1* binding sequences found in the corresponding genes of chicken and quail. The gene starts to work under 6.5 day of development (30 stage) and continue to 8.5 day of development (the stage 35). It is important that aromatase is not activated in males. It is also assumed that this enzyme regulates *AMH* activity in both sexes. In males the expression of *AMH* was found with 25 stages.

Human *DAX1* (Dosage sensitive sex reversal) is localized on the X chromosome. Its duplications are results sex reversion of the male-female. Gene expresses in the gonads of mice during sexual differentiation and acts as a kind of "antitestis" factor inhibiting *SRY*. The gene encodes a nuclear receptor with some unusual DNA-binding domain of the type of the repeat at the amino end. This chicken protein evinces 63% homology to human proteins, however, it does not have tandem repeats at the amino end. Gene is localized to the long arm of chromosome 1. It is more active during differentiation in embryonic gonads than in the testes. This gene is involved in steroidogenesis and gametogenesis in both sexes. *WT1* gene is active in both sexes. Genes that possess different activity in both sexes and a working scheme of sex determination are presented in the following figure (**Figure 8**) [6, 37].

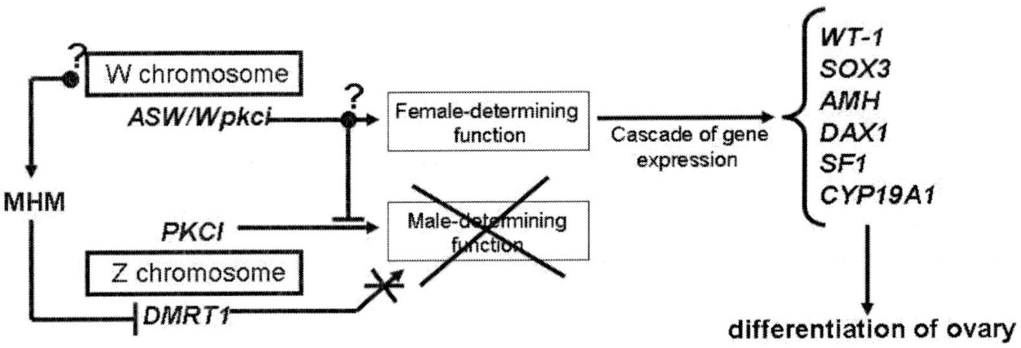

Figure 8. Working model of female sex determination in the chicken. Genes that play a role in sex determination in females are shown.

Sexual differentiation in birds is more conservative and somewhat different from that of mammals: 1) *SRY* gene is not present, although there are sex chromosomes, however, they are different from those in mammals; 2) Other genes that influence the development of the gonads are identified.

Gene *DMRT1* is strongly expressed in males in the early stages of development (stage 22) and can be of a dose-Z factor in the differentiation of sex. It is able to demonstrate the impact of using an aromatase inhibitor like fadrosole to ZW embryos that here as in ZZ individuals is a significant increase in the activity of the *DMRT1* gene compared with *SOX9* gene. It is assumed that the development of testes do not need two doses of *DMRT1*, no matter the difference in the dose and the activity of this gene is necessary for the differentiation of testes. Turning this gene by RNA interference in ZZ embryos leads to the feminization of the gonads and the evidence in favor of the hypothesis Z dose of sex determination in chickens. Interestingly a conserved region of 180bp *TESCO* (ECR-evolutionary conserved region) was described in the birds as well as the monotremes, marsupials, reptiles, amphibians. It described a conserved region of 180bp *TESCO* part of an enhancer responsible for the activation of gene *Sox9* gene in mammals. It is assumed that in vertebrates expression of *SOX9* governed through ECR with SOX, TCF/LEF, Forkhead, DMRT, GATA protein. For birds such role plays *DMRT1*. This gene is expressed in birds prior to *Sox9* and probably regulates its expression like *SRY* [38]. The appearance of testes still depends on AMH (antimullerian hormone). *AMH* gene expressed in both sexes at a low level but its activity is increased in males during differentiation of the gonads. It is assumed that its activation is independent of the *SOX9* gene. It is a very important question about the regulation of genes involved in the control of sex: as *DMRT1* and *AMH* genes are expressed in the gonads of both sexes, but with greater intensity in the males (**Figure 9**).

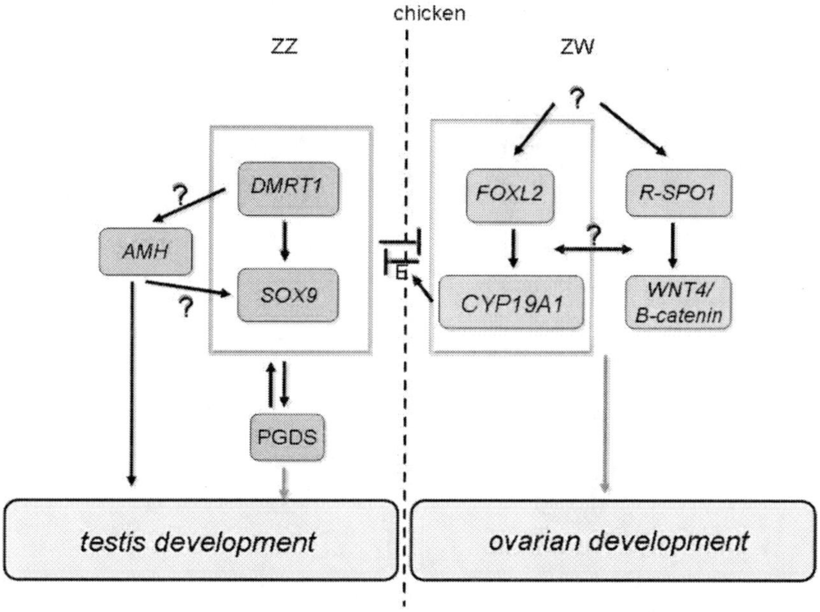

Figure 9. Genes are involved in the determination of testes and ovaries in chicken. The male *DMRT1* gene probably indirectly activates expression *SOX9* which is crucial for the emergence of testes jointly with *AMH.* In females *RSPO1* gene leads to activation *Wnt4/β-catenin* path and jointly with *FOXL2/aromatase* leads to differentiation of ovaries. *DMRT1* and *FOXL2* may exercise antagonizing differentiation testes and ovaries as is the case in mammals.

It is thought that the development of the ovaries in birds is difficult to adjust. First the transcription factor *FOXL2* works by controlling production of estradiol in target genes such as *CYP19A*, *FST*, *CEP11A*, *CYP17*, *cyclin D2*. Second the paths *FOXL2/aromatase*, *RSPO1*, *WNT4*, *β-catenin* are coordinated interaction in the development of the genital tract females. It also means participation *BMP* (bone morphogenetic proteins) path in the process [6, 21, 25].

Remain and different options for participation W chromosome (*HINTW*, *F factor*, etc.). There was a scheme of bird sex determination proposed where the W-encoded factors (*Factor F*) mediates the expression of Z chromosome *DMRT1* gene. The neighboring MHM area, containing 2.2 kb tandem repeat is hypermetilated in both chromosomes of males and it is hypometilated females in the same one female region of Z. W chromosome is required for hypometilation. In triploid 3A:ZZZ MHM region is not active and in 3A:ZZW individuals both Z transcribed as W is present. The following scheme of sex determination in birds through interaction W and Z chromosomes was proposed (**Figure 10**). RNA - seq was used to produce a comprehensive profile of gene expression in chicken blastoderm and embryonic gonads prior to sexual differentiation. The robust sexually dimorphic gene expression in both tissues pre-dating gonadogenesis including sex-linked and autosomal genes was found. These data support the hypothesis that sexual differentiation at the molecular level is at least partly cell autonomous in birds. The number of known genes of W chromosome expanded to 26. This include two novel W-linked sequence and three small RNAS reassigned to the chromosome. For virtually all expressed W-linked genes, the Z gametologue was also expressed and in both tissue. The exception was FAF (Female Associated Factor) which lacks a Z gametologue. There was high sequence homology between almost all W and Z linked copies at DNA level (average 88,4% identity) and protein level (average 90,3% identity). An exception was the gene *HINT-W* which showed 41% sequence and 48,5% amino acid homology. These results show a clear bias for genes associated with sex and reproduction is located on the avian Z sex chromosome making sex determination via Z dosage and expression level. For example Z-linked testis-associated genes *DMRT1* and *HEMOGEN* are more highly expressed in the gonads compared to the blastoderm. It is possible that additional Z-linked genes may play a role in sexual differentiation in different tissues through changes in their relative expression [40].

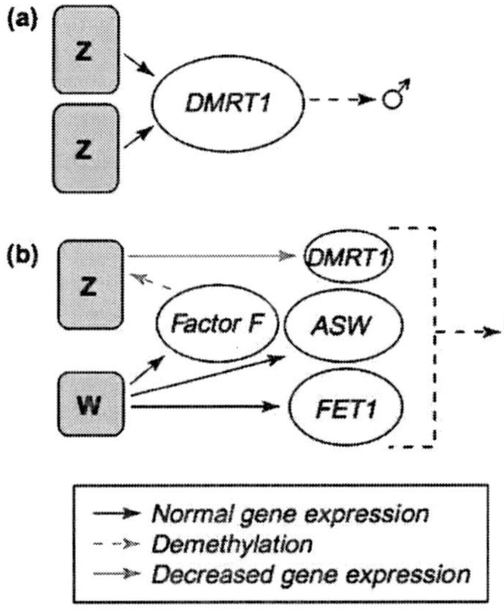

Figure 10. Potential mechanism of sex determination in birds. The mechanism uses a different dose of *DMRT1* gene in males (a) and females (b) and is mediated by an unknown factor F from W chromosome. W chromosome contains *ASW*, *FET1* gene [39].

So birds have got original version of genetical sex determination (GSD) implemented conservation certain key genes leaving the central role of estrogen characteristic of lower vertebrates, the most important functional interaction between the Z and W chromosomes, which do not have Y chromosomal *SRY* of mammal. Sex in chicken is determined by the sex chromosomes that must harbor one or more master sex determining genes. These master regulators control the expression and activity of several important downstream factors, including the male upregulating *DMRT1* and *SOX9*, and the female specific *FOXL2* and *RSPO1/WNT4* partways [6, 22, 40].

Sex Reversal in Bird

The female of 3-4 day of embryonic development develop testes or ovotestis after degenerative diseases of the ovaries, polyploidy, left ovariectomy, extraembryonic testis grafting, hypertrophy of the right gonads, implanted testes of 13 day-old etc. Interestingly the males in the chick embryos are heavier than females, and the anti-androgen injection on day 8 of incubation reduces the difference. Castration, transplantation, injection of estrogen, anti-estrogens, aromatase inhibitors can alter sex in birds, but the results are difficult to interpret in the sense of the primary sex determination [22]. Sex can change but in most cases the effect is temporary. Experimentally induced female-to-male sex reversal using the aromatase enzyme inhibitor fadrozole showed elevated levels of *DMRT1* expression similar to those of normal males (with two copies of the Z chromosome). It implies that the gene involvement in male development [41]. Aromatase inhibitor also reduced expression of *FOXL2* in the embryonic chicken ovary [42]. Although sex-reversed chickens developed testes containing essentially the same cellular components as those of normal male testes and were capable of complete spermatogenesis the limitations encountered are low sperm counts and blind-ended vas deferens production. However, as a rule, male fertility is not obtained. At least the number of sperm is not enough for a normal mating and fertilization. It is difficult to release sperm of the seminiferous tubules. There has been evidence indicating that the W chromosome containing sperm is functional, at least for fertilization. Usually only the right ovary is partly converted to in testis [43-46].

So, oestrogen obviously plays a key role in the gonadal differentiation in birds and genetically female chicken embryos can develop testes and a male phenotype if oestrogen syntesis is inhibited by treatment with an aromatase inhibitor before sexual differentiation. Sexual differentiation in birds is directed by oestrogen and exogenous compounds that interfere with oestrogen function may disrupt differentiation with regard to both the reproductive organs and brain [47].

The analysis of methylation status of chicken male hypermethylation region (MHM) and expression levels of sex dependent genes in adult female to male sex reversed chicken has shown that methylation of cMHM amplificon was lowest in the ovaries of standart females and highest in testes of standart males. Expression of *FOXL2*, *AMH*, *P450arom*, *SF1*, *ER* obviously higher in standard males, slightly and highly reversed female [48, 49].

A growing body of evidence reveals the importance of epigenetic regulatory mechanisms such as DNA-methylation, histone modifications and the role of non-coding RNAs in controlling sex determination and gonadogenesis. Chicken Z-linking MHM region and specific sites of hypermethylated inside chicken cyp19 involve [50]. The protocol was developed to feminize ZZ-male chicken embryonic gonads in a highly quantitative manner by direct injection of emulsified ethynylestradiol into yolk at various developmental stages. It was shown that a 170-nt region in the promoter of CYP19A1/aromatase, a key gene required for ovarian estrogen biosynthesis and feminization of chicken embryonic gonads contains highly quantitative, nucleotide base-level epigenetic marks that reflect phenotypic gonadal sex differentiation. This study suggests that the "epigenetic sex" of chicken embryonic gonads is more stable than the morphologically or transcriptionally characterized sex differentiation suggesting the importance of the nucleotide base-level "epigenetic sex" in gonadal sex differentiation [51]. The demethylating agent 5-aza-cytidine causes an asymmetric effect on the two Z chromosomes of males altering. The chromatin configuration, MHM RNA expression, and H4K16Ac modification suggest an inequality in the methylation status and chromatin of the two Z chromosomes [52].

Dosage Compensation in Birds and Sex Determination

There has not yet resolved the question of the presence of dosage compensation in birds Z chromosome genes in both sexes. On chicken Z chromosome is currently mapped about 1000 genes, 26 of which are also represented on the W chromosome. One would expect of a compensation mechanism for increased doses Z genes in males as these phenomena are described in mammals, fruit flies, nematodes. Initially, it was shown that the gene *ACO1* (aconitase) is expressed in ZZ male twice stronger than the ZW females which may indicate a lack of dosage compensation. It was also shown that the birds after the BrdU inclusion is not asynchronous replication between euchromatic disks ZZ pair in males, by analogy with a pair of XX mammals. In experiments using PCR, FISH and real-time PCR were showed that the ratio of the intensity of expression for Z chromosome genes in males and females at the transcriptional level varies from 0.8 to 2.4. For genes affected dosage compensation it should be about 1 and in the absence of this process about 2. It is suggested that only part of the Z chromosome genes are subject of compensation at least at the level of transcription. In chick the average gene expression Z chromosome in 1.4-1.6 times higher in males than in females. By using microarrays (microarray) in several tissues of the chicken and zebra finches, *Sylvia communis* was investigated relative expression at the transcriptional level. It was shown that sex chromosome dosage compensation surprisingly ineffective in birds at least in comparison with mammals. Higher expression of some Z chromosome genes in males is associated with their possible role in the control of sexual differentiation. There are about 20 of these genes. Only 30-40% of the genes of chromosome Z more efficiently expressed in males than in females [53, 54].

In birds possible mechanism of dosage compensation is described by M. Teranishi and coauthors [55] in which the identified site in Z chromosome - MHM (male hypermethylated region) next to the *DMRT1* gene which contains about 200 copies of the sequence length of 2 Mbp. This area also contains hypermethylated CpG islands in genomes males and intersexes, starting from 1 day of embryonic development. Transcription of high molecular 10 kb polyA mRNA observed with these nucleotide units only in a single Z chromosome and a single strand of DNA in females. This RNA has an open reading frame, it is synthesized in the cells of females, remains in the nucleus and accumulates near MHM or *DMRT1* locus. It may be one of the factors causing the silence *DMRT1* gene in females after hatching (**Figure 11**), [10].

A) Hypothesis Hypermethylation in Males (CMA Hypothesis)

CMA region in ZZ embryos is transcriptionally silent due to hypermethylation. Same male determining neighboring genes such as *DMRT1* free for transcription. In some embryos

ZW having W chromosome genes (Factor F) induce demethylation of MHM providing transcription high molecular noncoding RNA which accumulates in the site of transcription. Transcription of neighboring genes such as *DMRT1* decreases. It is assumed that the CMA and hypomethylation *DMRT1* transcription repression of females are under the influence of W chromosome. So sex determination is associated with active transcription of the gene *DMRT1* in males compared with females. Thus manifested a marked resemblance polyA mRNA MHM with Xist and roX1, roX2 RNA in human and fruit fly *(Drosophila melanogaster)* but the spreading MHM limited. In the latter case untranslated transcripts also raise and regulate certain compensating dose RNP complexes. It should be noted that this similarity outside and there is no direct evidence of the contribution of MHM in compensation dose [56].

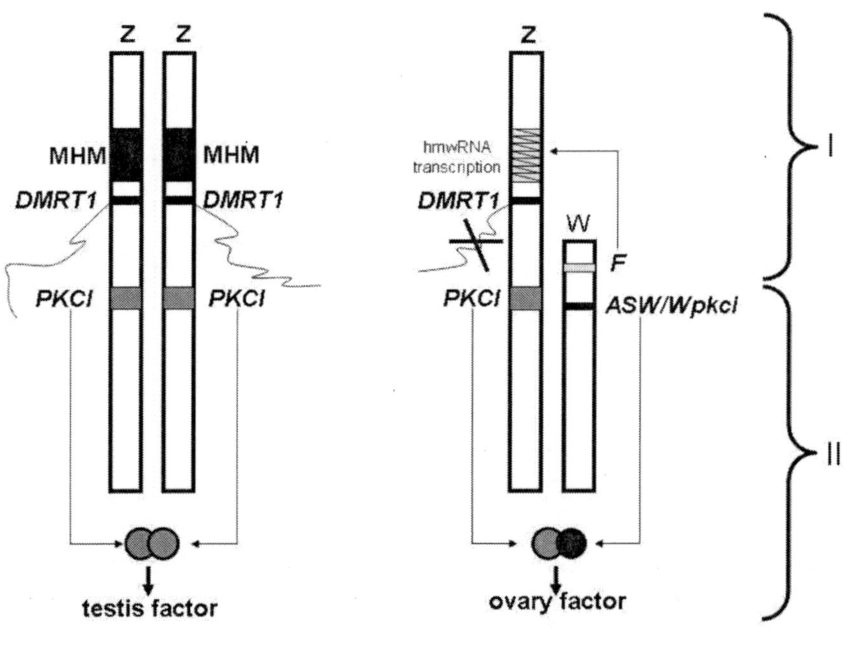

Figure 11. Models of sex determination in birds involving interaction Z and W chromosomes.

It is important to note that the MHM coincides with the cytogenetic position of *DMRT1* on Z chromosome which may allow the gene to work actively in males. Articles of Melamed E. and Arnold A. [53] assessment of regional differences in dosage compensation on the Z chromosome demonstrate that it is in part Zp (Z valley) localized cluster of compensated genes. According to other data, there are no areas of dosage compensation on the Z chromosome and the corresponding genes are regulated individually depending on the tissue and the stage of ontogenesis. It is assumed that 45%, 38% and 12% Z-chromosome genes respectively compensated, non-compensated or partially compensated. The role of the W chromosome may be to control hypomethylation and transcriptional activity of the corresponding section of Z chromosome. In intersexes ZZW observed hypomethylation and transcription MHM, and males ZZZ-hypermethylation and lack of transcription. It is interesting that the zebra finch differs from the chicken because it lacks a specialized region of greater dosage compensation along the Z chromosome and shows other differences in sex bias. These patterns suggest that different avian taxa may have evolved specific compensatory mechanisms [57].

Modification of Z chromosome in the area Zp21 of MHM localization was described. This area is strongly enriched in chromatin acetylated at lysine 16 histone H4 in the chromosome of females but not males [56]. The same specific modification of histone observed in the X chromosome in *Drosophila* males throughout its length in the process of compensation in this species. There is an intriguing parallel between dosage compensation in *Drosophila* and birds. It is assumed that the birds have a partial and variable gene dosage compensation. CR1-LINE elements are more represented in the chicken Z chromosome than in the autosomes but apparently not directly related to dosage compensation. On the contrary CpG-islands are centered around uncompensated genes and can influence it. It should be noted that the *Zebra amaziny* as the second bird with sequenced genome has not got a specialized area with a large dosage compensation in contrast to the chicken Z chromosome. In this species there is a reduction male/female ratio in respect of gene expression near MHM. There is no here specific sequence and acetylation H4K16. It was shown that the MHM locus evinces configuration differences in males (condensed) and females (open) and identified numerous MHC sequence and RNA.

This suggests that the epigenetic sex of chicken embryonic gonads is more stable than the morphologically or transcriptionally characterized sex differentiation, suggesting the importance of the nucleotide base-level epigenetic sex in gonadal sex differentiation. Bannister *et al.* [57] reported the identification of miRNA202*(miR-202*) with male-biased expression and hypothesized that it could be implicated in testicular differentiation. To address this hypothesis, they induced feminization by the injection of eggs at embryonic day 4.5 (E4.5) with E2 and analyzed changes in miR-202*expression which was reduced to female levels and correlated with reduced expression of the testis-associated genes *DMRT1* and *SOX9*, and with up-regulation of the ovary-associated genes *FOXL2* and *CYP19A*. On the other hand, female gonads treated at E3.5 with aromatase inhibitor which blocks estrogen synthesis were masculinized by E9.5. In this case miR-202* expression was increased and correlated with down-regulation of *FOXL2* and *CYP19A* and up-regulation of *DMRT1* and *SOX9*. .Thus up-regulation of miR-202* coincides with testicular differentiation in embryonic chicken gonads. Sex marks in the *CYP19A1/aromatase* promoter involving DNA methylation and histone lysine methylation are feminized significantly but only partially in sex-converted gonad seven when morphological and transcriptional marks of sex differentiations how complete feminization on, being in distinguishable from gonads of normal ZW females.

MHM assay have used for bird molecular sexing. All males showed hypermethylation of the MHM region and all females showed - hypomethylation [60]. The down-regulation of *DMRT1* and up-regulation of *ERa* in rooster testes treated with exogenous pEGFP-N1-cMHMplasmid might indicate a novel cMHM non-coding RNA-mediated mechanism in avian sex differentiation [61].

In platurus already appears dosage compensation involving the transcription of only one of the two alleles in some of the cells and then compensated with features such Z chromosome of birds and mammals (the X chromosome). Interestingly *Palaeognathae* Z and W are practically the same and there are only the initial stages of the evolution of compensation dose. This is, in particular, in the presence of the Z and W chromosomes of two genes *Ratitae ACO1* and *ZOV3* [62].

In birds, unlike mammals, described a special cell-autonomous mechanism of gender identity (CASI-sex autonomous sex identity). Cell autonomous sexual differentiation would involve purely intrinsic genetic factors independent of any extrinsic signalling. There are the data that show sexually dimorphic gene expression in chicken blastoderms and gonads prior to gonadal sex differentiation. This supports the notion that sexual differentiation begins at the molecular level in a cell autonomous manner in the chicken embryo [6, 63]. It is assumed that both CASI and hormones affect the final tissue phenotype. Genes, they are designed CASI effect in chickens is expected about 300. It is assumed that in insects, crustaceans and even marsupials also marked the phenomenon of sexual identity inheritance cells [64, 65].

Conclusion

The mammalian male-determining factor *SRY* does not exist in birds. Another transcription factor *DMRT1* plays control role in bird testis development. Ovarian differentiation is less well characterized. There are several important factors regulating ovarian development and studies in bird embryos. Sex in chickens is determined by the sex chromosomes that must harbor one or more master sex determining genes. These master regulators control the expression and activity of several important downstream factors including the male up regulated *DMRT1* and *SOX9*, and the female-specific *FOXL2* and *RSPO1/WNT4* pathways.

References

[1] Gamble, T., Zarkower, D. (2012). Sex determination. *Curr. Biol.*, 22 (8): R257-262.

[2] Nanda, I., Haaf, T., Schartl, M., Schmid, M., Burt, DW (2002). Comparative mapping of Z-orthologous genes in vertebrates: implications for the evolution of avian sex chromosomes. *Cytogenet Genome Res.*, 99:178-184.

[3] Shetty, S., Griffin, D. K., Graves, J. A. M. (1999). Comparative painting reveals strong chromosome homology over 80 million years of bird evolution. *Chromosome Res.*, 7: 289–295.

[4] Ellegren, H. (2011). Emergence of male-biased genes on the chicken Z-chromosome: Contrasts between male and female heterogametic systems. *Genome Research.*, 21:2082-2086.

[5] Alekseevich, L.A., Lukina, N.A., Nikitin, N.S., Nekrasova, A.A., Smirnov, A.F. (2009). Problems of sex determination in birds exemplified by Gallus gallus domesticus. *Russian Journal of Genetics,* 45(3):255-265.

[6] Ayers, K. L., Smith, C. A., Lambeth, L. S. (2013). The Molecular Genetics of Avian Sex Determination and its Manipulation. *Genesis*, 51:325–336.

[7] Chen, N, Bellott, D.W., Page, D.C., Clark, A.G. (2012). Identification of avian W-linked contigs by short-read sequencing. *BMC Genomics*, 13:183.

[8] Consortium ICG: Gallus gallus 4.0. 22/11/2011 edition; 2011.

[9] Rodrigue, K.L., May, B.P., Famula, T.R., Delany, M.E. (2005). Meiotic instability of chicken ultra-long telomeres and mapping of a 2.8 megabase array to the W-sex chromosome. *Chromosome Res*, 13(6):581-591.

[10] Smith, C.A., Sinclair, A.H: (2004). Sex determination: insights from the chicken. *Bioessays*, 26(2):120-132.

[11] Stiglec, R., Ezaz, T., Graves, J.A. (2007). A new look at the evolution of avian sex chromosomes. *Cytogenet Genome Res*, 117(1-4):103-109.

[12] Waters, P.D., Graves, M.J.A. (2009). Monotreme sex chromosomes implications for the evolution of amniote sex chromosomes. *Reproduction, fertility and development,* 21: 943-951.

[13] Nam, K, Ellegren, H. (2008). The chicken (Gallus gallus) Z chromosome contains at least three nonlinear evolutionary strata. *Genetics*, 180(2):1131-1136.

[14] Wright, A.E., Moghadam, H.K., Mank, J.E. (2012). Trade-off between selection for dosage compensation and masculinization on the avian Z chromosome. *Genetics*, 192(4):1433-1445.

[15] Vicoso, B, Kaiser, V.B., Bachtrog, D. (2013). Sex-biased gene expression at homomorphic sex chromosomes in emus and its implication for sex chromosome evolution. *Proc Natl Acad Sci USA*. 110(16):6453-6458.

[16] Clinton, M, Haines, L.C. (2001). An overview of factors influencing sex determination and gonadal development in birds. *EXS*.(91):97-115.

[17] Graves, J.A.M. (2003). Sex and death in birds: A model of dosage compensation that predicts lethality of sex chromosome aneuploids. *Cytogenet Genome Res,* 101:278–282.

[18] Lambeth, L.S, Smith, C.A. (2012). Disorders of sexual development in poultry. *Sex Dev*, 6(1-3):96-103.

[19] Küpper, C., Augustin, J., Edwards, S., Székely, T., Kosztolányi, A., Burke, T., Janes, D. E. (2012). Triploid plover female provides support for a role of the W chromosome in avian sex determination. Biol. Lett, 8 (5):787-789.

[20] Clinton M. (1998). Sex determination and gonadal development: A bird's eye view. *Journal of Experimental Zoology*, 281(5):457–465.

[21] Ayers, K.L., Sinclair, A.H., Smith, C.A. (2013). The molecular genetics of ovarian differentiation in the avian model. *Sexual Development*, 7 (1-3):80-94.

[22] Ditewig, A.C., Yao, H.H. (2005). Organogenesis of the ovary: a comparative review on vertebrate ovary formation. Organogenesis, 2(2):36-41.

[23] Pask, A.J. (2012). A role for estrogen in somatic cell fate of the mammalian gonad // *Chromosome Res*, 20(1):239–45.

[24] Trukhina, A.V., Lukina, N.A., Wackerow-Kouzova, N.D., Smirnov A.F. (2013). The variety of vertebrate mechanisms of sex determination. Current Advances in Molecular Phylogenetics, in press.

[25] Cutting, A, Chue, J, Smith, CA. (2013). Just how conserved is vertebrate sex determination? // *Dev Dyn*, 242(4):380-387.

[26] Adkins-Regan, E. (2012). Hormonal organization and activation: evolutionary implications and questions. *Gen Comp Endocrinol*, 176(3):279-285.

[27] Fridolfsson, A.K., Ellegren, H. (2000). Molecular evolution of the avian CHD1 genes on the Z and W sex chromosomes. *Genetics*, 155(4):1903-1912.

[28] Pace, H. C., Brenner, C. (2003). Feminizing chicks: a model for avian sex determination based on titration of Hint enzyme activity and the predicted structure of an Asw-Hint heterodimer. *Genome Biology*, 4:R18.

[29] Smith, C.A., Roeszler, K.N., Hudson, Q.J,, Sinclair, A.H. (2007). Avian sex determination: what, when and where? *Cytogenet Genome Res*, 117(1-4):165-173.

[30] Smith, C. A., Roeszler, K. N., Sinclair, A. H. (2009). Genetic evidence against a role for W-linked histidine triad nucleotide binding protein *(HINTW)* in avian sex determination. *Int. J. Dev. Biol*, 53: 59 – 67.

[31] Reed, K.J., Sinclair, A.H. (2002). FET-1: a novel W-linked, female specific gene up-regulated in the embryonic chicken ovary. *Mech Dev*, 119(1):S87-90.

[32] Cutting, A.D., Bannister, S.C., Doran, T.J., Sinclair, A.H., Tizard, M.V., Smith, C.A. (2012). The potential role of microRNAs in regulating gonadal sex differentiation in the chicken embryo. *Chromosome Res,* 20(1):201-213.

[33] Smith, C.A., Roeszler, K.N., Ohnesorg, T., Cummins, D.M., Farlie, P.G., Doran, T.J., Sinclair, A.H. (2009). The avian Z-linked gene DMRT1 is required for male sex determination in the chicken. *Nature*, 461 (7261):267-271.

[34] Smith, C. A. (2010). Sex determination in birds: a review. *Emu*, 110(4):364-377.

[35] Chue, J, Smith, C.A. (2011). Sex determination and sexual differentiation in the avian model. *FEBS J,* 278(7):1027-1034.

[36] Shetty, S., Kirby, P, Zarkower, D., Graves, J.A. (2002). DMRT1 in a ratite bird: evidence for a role in sex determination and discovery of a putative regulatory element. *Cytogenet Genome Res,* 99(1-4):245-251.

[37] Mizuno, S, Kunita, R, Nakabayashi, O, Kuroda, Y, Arai, N, Harata, M, Ogawa, A, Itoh, Y, Teranishi, M, Hori, T. (2002). Z and W chromosomes of chickens: studies on their gene functions in sex determination and sex differentiation. *Cytogenet Genome Res*, 99(1-4):236-244.

[38] Bagheri-Fam, S, Sinclair, A.H, Koopman, P, Harley, V.R. (2010). Conserved regulatory modules in the Sox9 testis-specific enhancer predict roles for SOX, TCF/LEF, Forkhead, DMRT, and GATA proteins in vertebrate sex determination. *Int J Biochem Cell Biol*, 42(3):472-477.

[39] Nakagawa, S. (2004). Is avian sex determination unique?: clues from a warbler and from chickens. *Trends in Genetics*, 20(10):479-480.

[40] Ayers, K.L., Davidson, N.M., Demiyah, D., Roeszler, K.N., Grutzner, F, Sinclair, A.H., Oshlack, A, Smith, C.A. (2013). RNA sequencing reveals sexually dimorphic gene expression before gonadal differentiation in chicken embryos and allows comprehensive annotation of W-chromosome genes. *Genome Biol,* 25;14(3):R26.

[41] Smith, C.A., Katz, M., Sinclair, A.H. (2003). DMRT1 is upregulated in the gonads during female-to-male sex reversal in ZW chicken embryos. *Biol Reprod*, 68(2):560-570.

[42] Hudson, Q.J/, Smith, C.A., Sinclair, A.H. (2005). Aromatase inhibition reduces expression of FOXL2 in the embryonic chicken ovary. *Dev Dyn,* 233(3):1052-1055.

[43] Elbrecht, A, Smith, R.G. (1992). Aromatase enzyme activity and sex determination in chickens. *Science*, 255(5043):467-470.

[44] Trefil, P., Micakova, A., Mucksova, J., Poplstein, M., Brillard, J. P., Hodek, P. (2004). Sex reversed chicks Gallus domesticus hatched from eggs treated with aromatase inhibitor YM511. *Czech Journal of Animal Science*, 49(12): 511-516.

[45] Vaillant, S, Guémené, D, Dorizzi, M, Pieau, C, Richard-Mercier, N, Brillard, J.P. (2003). Degree of sex reversal as related to plasma steroid levels in genetic female chickens (Gallus domesticus) treated with Fadrozole. *Mol Reprod Dev*, 65(4):420-428.

[46] Shimada, K, Valdez, M.B. Jr, Mizutani, M, Namikawa, T. (2007). Potential application of sperm bearing female-specific chromosome in chickens. *Cytogenet Genome Res*,117(1-4):240-247.

[47] Brunström, B, Axelsson, J, Halldin, K. (2003). Effects of endocrine modulators on sex differentiation in birds. *Ecotoxicology*, 12(1-4):287-295.

[48] Yang, X., Zheng, J., Na, R., Li, J., Xu, G., Qu, L., Yang, N. (2008). Degree of sex differentiation of genetic female chicken treated with different doses of an aromatase inhibitor. *Sex Dev*, 2(6):309-315.

[49] Yang, X., Zheng, J., Qu, L., Chen, S., Li, J., Xu, G., Yang, N. (2011). Methylation status of cMHM and expression of sex specific genes in adult sex-reversed female chickens. *Sexual Development*, 5 (3):147-154.

[50] Piferrer, F. (2013). Epigenetics of sex determination and gonadogenesis. *Dev Dyn*, 242(4):360-370.

[51] Ellis, H.L., Shioda, K., Rosenthal, N.F., Coser, K.R., Shioda, T. (2012). Masculine epigenetic sex marks of the CYP19A1/Aromatase promoter in genetically male chicken embryonic gonads are resistant to estrogen –induced phenotypic sex conversion. *Biology of Reproduction*, 87(1):23, 1-12.

[52] Itoh, Y, Kampf, K, Arnold, A.P. (2011). Possible differences in the two Z chromosomes in male chickens and evolution of MHM sequences in Galliformes. *Chromosoma*, 120(6):587-598.

[53] Melamed, E, Arnold, A.P. (2007). Regional differences in dosage compensation on the chicken Z chromosome. *Genome Biol,* 8(9):R202.

[54] Zhang, S.O., Mathur, S., Hattem, G., Tassy, O., Pourquié, O. (2010). Sex-dimorphic gene expression and ineffective dosage compensation of Z-linked genes in gastrulating chicken embryos. *BMC Genomics*, 11:13.

[55] Teranishi, M., Shimada, Y., Hori, T., Nakabayashi, O., Kikuchi, T., Macleod T., Pym, R., Sheldon, B., Solovei, I., Macgregor, H., Mizuno S. (2001). Transcripts of the MHM region on the chicken Z chromosome accumulate as non-coding RNA in the nucleus of female cells adjacent to the DMRT1 locus. *Chromosome Res*, 9(2):147-165.

[56] Roeszler, K.N., Itman, C., Sinclair, A.H., Smith, C.A. (2012). The long non-coding RNA, MHM, plays a role in chicken embryonic development, including gonadogenesis. *Dev Biol*, 15,366(2):317-326.

[57] Itoh, Y, Replogle, K, Kim, Y.H., Wade, J., Clayton, D.F., Arnold, A.P. (2010). Sex bias and dosage compensation in the zebra finch versus chicken genomes: general and specialized patterns among birds. *Genome Res,* 20(4):512-518.

[58] Bisoni, L., Batlle-Morera, L., Bird, A.P., Suzuki, M., McQueen, H.A. (2005). Female-specific hyperacetylation of histone H4 in the chicken Z chromosome. *Chromosome Res*, 13(2):205-214.

[59] Bannister, S.C., Smith, C.A., Roeszler, K.N., Doran, T.J., Sinclair, A.H., Tizard, M.L. (2011). Manipulation of estrogen synthesis alters MIR202* expression in embryonic chicken gonads. *Biol Reprod*, 85(1):22-30.

[60] Caetano, L.C., Ramos, E.S. (2008). MHM assay: molecular sexing based on the sex-specific methylation pattern of the MHM region in chickens. *Conservation Genetics*, 9, I. 4:985-987.

[61] Yang, X, Zheng, J, Xu, G, Qu, L, Chen, S, Li, J, Yang, N. (2010). Exogenous cMHM regulates the expression of DMRT1 and ER alpha in avian testes. *Mol Biol Rep,* 37(4):1841-1847.

[62] Livernois, A.M., Graves, J.A., Waters, P.D. (2012). The origin and evolution of vertebrate sex chromosomes and dosage compensation. *Heredity (Edinb)*, 108(1):50-58.

[63] Arnold, A.P., Chen, X., Link, J.C., Itoh, Y., Reue, K. (2013). Cell-autonomous sex determination outside of the gonad. *Dev Dyn,* 242(4):371-379.

[64] Zhao, D., McBride, D., Nandi, S., McQueen, H.A., McGrew, M.J., Hocking, P.M., Lewis, P.D., Sang, H.M., Clinton M. (2010). Somatic sex identity is cell autonomous in the chicken. *Nature*, 464:237-242.

[65] Clinton, M., Zhao, D., Nandi, S., McBride, D.. (2012). Evidence for avian cell autonomous sex identity (CASI) and implications for the sex-determination process? *Chromosome Res,* 20(1):177-190.

In: Birds
Editors: Lucas Ruiz and Franco Iglesias
ISBN: 978-1-62948-104-3

Chapter 2

Structure and Function of the Avian Respiratory System, with Observation on its Predisposition to Injury by Particulates and Pathogenic Microorganisms

J. N. Maina*
Department of Zoology, University of Johannesburg, Johannesburg, South Africa

Abstract

Among the air-breathing vertebrates, the avian respiratory system, the lung-air sac system, is structurally the most complex and functionally the most efficient gas exchanger. The lung which is small and compact is firmly held between the ribs and the vertebrae while the air sacs are capacious transparent membranous structures which are widely spread in the coelomic cavity. The lung is ventilated unidirectionally and continuously by a bellows-like synchronized activity of the air sacs. It takes two inspiratory cycles and two exhalatory cycles for the air to enter and exit the lung. Morphometrically, the avian lung is highly specialized. Compared to a mammal of equivalent body mass, the bird lung has a respiratory surface area which is 15% greater and the blood-gas barrier is 62% thinner. Free (surface) macrophages are scarce on the respiratory surface of the avian lung. Because of this it has been argued that birds have poor pulmonary cellular defences and are therefore highly vulnerable to injury by particulates and infection by pathogens. This conclusion has, however, been questioned. Diverse phagocytic cells which include subepithelial macrophages, pulmonary intravascular macrophages, bronchial epithelial cells, and free macrophages should provide adequate protection. The large tidal volume of birds and the extension of the air sacs to lie superficially, are some of the features which make birds susceptible to pulmonary afflictions and infections.

* E-mail: jmaina@uj.ac.za.

Definition of Abbreviations

AAS(s)	Abdominal Air Sac/Abdominal Air Sacs
AC(s)	Air Capillary/Air Capillaries
AS(s)	Air Sac/Air Sac(s)
BC(s)	Blood Capillary/Blood Capillaries
BGB	Blood-Gas Barrier
EPPB	Extrapulmonary Primary Bronchus
ET	Exchange Tissue
FM(s)	Free Macrophage/Free Macrophages
IAV	Inspiratory Aerodynamic Valving
IPPB	Intrapulmonary Primary Bronchus
IPRS	Interparapronchial Septum/ Interparapronchial Septa
LDSB	Laterodorsal Secondary Bronchus/Laterodorsal Secondary Bronchi
LVSB	Lateroventral Secondary Bronchus/Lateroventral Secondary Bronchi
MDSB	Mediodorsal Secondary Bronchus/Mediodorsal Secondary Bronchi
MVSB	Medioventral Secondary Bronchus/Medioventral Secondary Bronchi
NPPR	Neopulmonary Parabronchi
PB	Primary Bronchus
PPPR	Paleopulmonic Parabronchi
PR	Parabronchus/Parabronchi
RSA	Respiratory Surface Area
VL	Volume of the Lung

Introduction

All modern birds, including those like the penguins, the rhea, the kiwi, the ostrich, the cassowaries, and the emu which are now flightless, evolved from flying progenitors (Welty, 1979). In various ways which relate to their lifestyle, birds are an exceptional taxon. Powered flight is a highly energetically demanding form of locomotion (Tobalske et al., 2003). Among the evolved animal taxa, it has occurred only in two phyla, namely the Chordata and the Arthropoda. Volancy chronologically evolved in insects ~350 million years ago (mya) (Wigglesworth, 1972), in the now extinct pterodactyls ~220 mya (Bramwell, 1971), in birds ~150 mya (De Beer 1954), and in bats ~50 mya (Yalden and Morris, 1975). During flight, the oxygen consumption (VO_2) of a bird increases 20 to 30 times the one at rest (Tucker, 1970). The high energetic demand for active flight has compelled remarkable morphological uniformity in birds, a taxon which comprises of ~9,000 species (Gruson, 1976). The external morphology of birds is more uniform than that in a single order of fishes, amphibians, and reptiles (Marshall, 1962). There is less difference in the external forms of birds than there is the 90 species of primates and 290 species of carnivores (Yapp, 1970). This shows that flight has been greatly prescriptive in setting optimal aerodynamic configuration of the external form of birds. In spite of the external uniformity, interestingly, remarkable variation occurs in the morphologies of the internal organs of birds, including those of the respiratory system (King and King, 1979).

A considerable metabolic barrier separates volant- from nonvolant animals. Subsequent to their independently evolving from reptiles long after mammals (Ostrom, 1975) and realizing endothermic homiothermy, birds attained metabolic capacities between rest and maximum exercise or cold induced thermogenesis which were 4 to 15 times higher than those of their antecedents at same body temperature (Lasiewski, 1962). The evolution of bird flight and with it the high energetic demands which accompany it was incontrovertibly underpinned by development of an exceptionally efficient respiratory system, the lung-air sac system. Its structure and function was recently reviewed by Maina (2005, 2006, 2008). Here, only a succinct account, mainly based on the respiratory system of the domestic fowl, *Gallus gallus* variant *domesticus*, is given. Additional details can be acquired from the previous comprehensive treatises.

Plausible morphological and physiological features which make the lung-air sac system susceptible to attack by pathogetic microorganisms and damage by inhaled toxicants are pointed out.

Structure of the Respiratory System

Lung

Among the gas exchangers which have evolved in the air-breathing vertebrates, the structure of the avian lung is outstanding. The lung is located on the dorsal aspect of the coelomic cavity where it is firmly attached to the vertebrae and the ribs (Figures 1-4). Between ⅕ to ⅓ of its volume is held between the ribs (King, 1966; Ducker, 1971; King and Molony, 1971). With the diaphragm lacking, the liver and not the lungs, as is the case in mammals, borders the heart.

The right- and left lungs, which are of equivalent volumes, are small (Figures 1, 2) and compact (Figures 5, 6) and unlike those of most mammals, it is never lobed. Wedge-shaped (Figures 3, 5, 6), the dorsolateral surface which contacts the ribs is called 'costal surface'; the dorsomedial side which contacts the vertebrae is designated 'vertebral surface', and; the ventromedial part which contacts the horizontal septum is termed 'septal surface'. The ostia, the openings which connect the lungs to the air sacs (ASs) are found on the septal surface. The hilum (Figures 3, 5) is the area where the primary bronchus (PB) and the pulmonary blood vessels and the nerves enter or leave the lung: it is located about the junction of the cranial- and middle thirds of the lung on the medial surface (Figures 3, 5, 7). The trachea bifurcates at the syrinx to form the left and right extrapulmonary primary bronchi (EPPB) (Figures 3) which, forming the intrapulmonary primary bronchi (IPPB), passes through the lung in a craniocaudal direction (Figure 7). As it passes through the lung, the IPPB give rise to four sets of secondary bronchi, namely the medioventral secondary bronchi (MVSB), the mediodorsal secondary bronchi (MDSB), the laterodorsal secondary bronchi (LDSB), and the lateroventral secondary bronchi (LVSB) (Figure 7). Hoop-like in form, the parabronchi (tertiary bronchi) (PR) interconnect the secondary bronchi (Figures 5, 7) and anastomose with each other (Figure 8). The area where the PR connect the MVSB to the MDSB is called 'planum anastomoticum' (Figure 9).

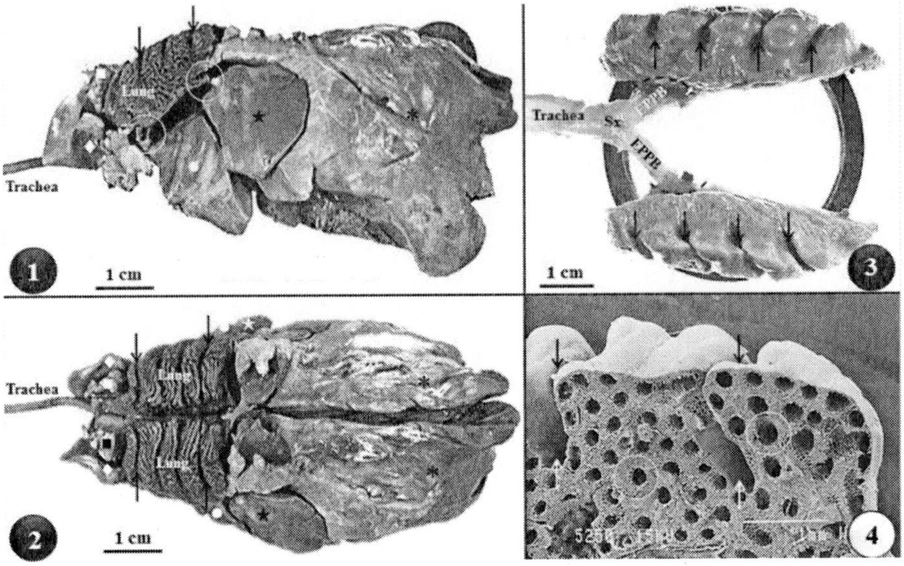

Figures 1.2. Latex rubber cast preparations of the lung and the air sacs of the domestic fowl, *Gallus gallus* variant *domesticus*. Arrows, costal sulci (insertions of the lung to the ribs and the vertebrae); circles, ostia (connections of the lung to the air sacs); square, cervical air sac; diamond, interclavicular air sac; spot, craniothoracic air sac; star, caudothoracic air sac; asterisk, abdominal air sac. Figures 3, 4. Insertions of lungs of the ostrich and the domestic fowl respectively to the ribs and the vertebrae at the costal sulci (arrows). Sx, location of the syrinx; EPPB, extrapulmonary primary bronchus; dashed line, hili; circles, parabronchi.

Primary Bronchus

After penetrating the horizontal septum, the EPPB enters and transits the lung as the IPPB. It passes through the lung in an arched manner and progressively decreases in diameter (King, 1966) (Figure 7). The IPPB terminates on the caudal (septal) margin of the lung by entering the abdominal air sac. Both the EPPB and the IPPB are supported by initially cartilaginous and later osseous plates which are enclosed in their walls. The lumina of the EPPB and first part of the IPPB are lined by a pseudotratified ciliated columnar epithelium which contains mucous secreting goblet cells.

Secondary Bronchi

Four sets of secondary bronchi which originate from the IPPB (Figure 7). They are named according to the parts of the lung they supply air to. The MVSB originate from from the dorsomedial aspect of the cranial one-third of the IPPB and the MDSB, the LVSB, and the LDSB originate from the caudal two-thirds of the it. The MDSB branch from the dorsal circumferential aspect of the lumen of the IPPB. Among the secondary bronchi, the LVSB differ considerably both in number and size: they originate in sequence from the IPPB at the same craniocaudal length of the IPPB as the MDSB but from the opposite aspect of the lumen of the airway, i.e., on the ventral surface of the IPPB, diametrically facing the openings of the MDSB.

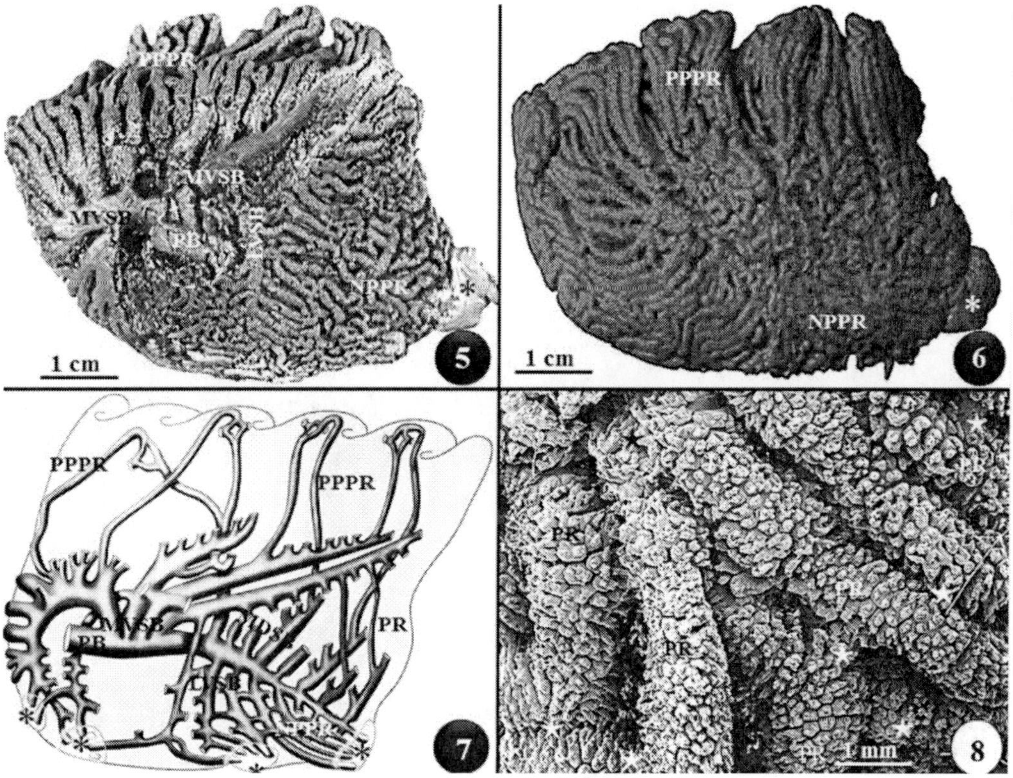

Figures 5-8. Latex cast (Figures 5, 6) and a diagram (Figure 7) of the air ways of the lung of the domestic fowl: Figure 5 is a medial view- and Figure 6 a lateral one. The airways include the primary bronchus (PB), the medioventral secondary bronchi (MVSB), the lateroventral secondary bronchi (LVSB), the paleopulmonic parabronchi (PPPR), and neopulmonic parabronchi NPPR). Asterisk, ostium to the abdominal air sac; circle, hilum, point where the primary bronchus enters the lung. Figure 7: The lung drawn as if it is transparent to show the arrangement of the airways. PB, primary bronchus; MVSB, medioventral secondary bronchi; MDSB, mediodorsal secondary bronchi, LVSB, lateroventral secondary bronchi; asterisk, ostia; PPPR, paleopulmonic parabronchi; NPPR, neopulmonic parabronchi. The laterodorsal secondary bronchi have been left out to simplify the diagram. Figure 8: Latex cast of the parabronchi (PR) showing sites where they interconnect (stars). The atria which project from the parabronchial lumen can be clearly seen as swellings on the parabronchi.

The LDSB stem from the same craniocaudal length of the IPPB as the MDSB and the LDSB but from the lateral wall of the IPPB: they are plentiful and small in diameter and structurally resemble the PR.

In contrast to the other groups of secondary bronchi which branch from the IPPB in succession, the LDSB have a disseminated origin and are highly variable in number and size. A short distance from the IPPB, the secondary bronchi are lined by an epithelium similar to that of the IPPB, i.e., a pseudotratified ciliated columnar epithelium with mucus secreting goblet cells: mucus glands are absent in the epithelium which lines the rest of the secondary bronchi (Duncker, 1971).

Most of the luminal surfaces of the secondary bronchus are lined by a cuboidal (sometimes ciliated) and squamous epithelium which is similar to that of the PR. The transition between of the epithelium which lines the secondary bronchi and that lining the PR may be abrupt (Del Corral, 1995).

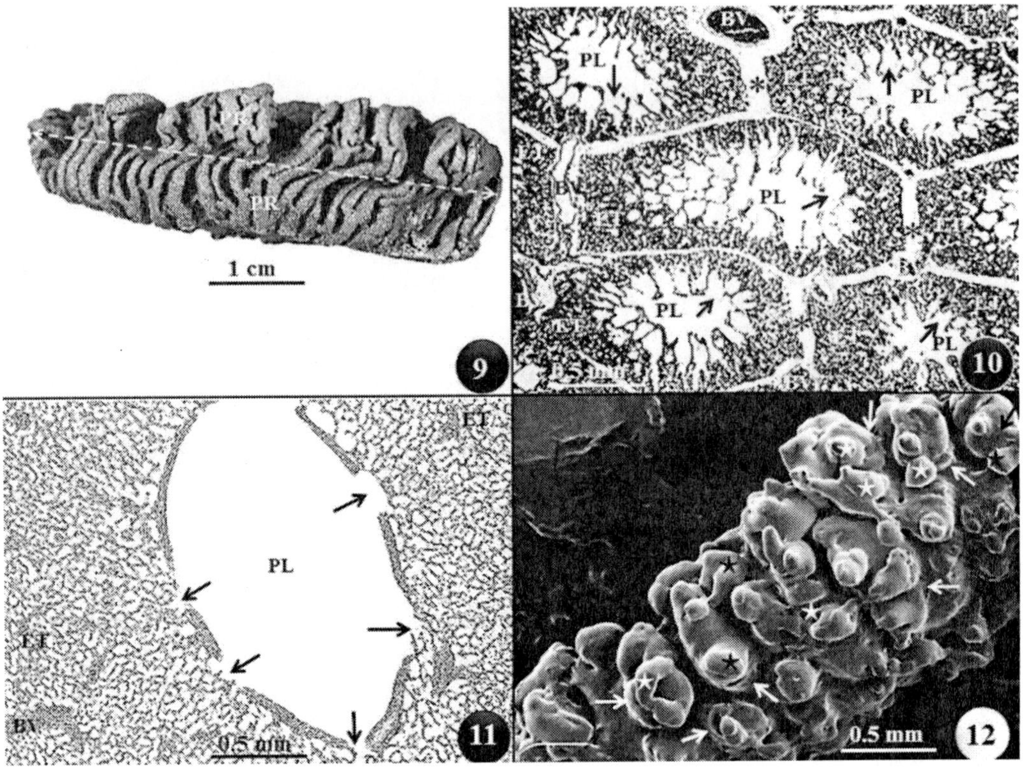

Figures 9-12. Figure 9: Dorsal view of the lung of the domestic fowl showing the linea anastomosticum (dashed line) area where parabronchi (PR) from the mediovetral secondary bronchi and those from the mediodorsal secondary bronchi connect. Figures 10, 11. Cross-sections of the parabronchi of the lung of the domestic fowl and the ostrich respectively. The parabronchi comprise of a parabronchial lumen (PL) which is surrounded by exchange tissue (ET). BV, interparabronchial blood vessel; asterisk, interparabronchial septa; arrows, atria which project from the parabronchial lumen. Figure 11. Latex rubber cast of a parabronchus showing atria (arrows) which project from the parabronchial lumen. They give rise to infundibulae (stars).

Parabronchi (Tertiary Bronchi) (PR)

In the lungs of the galliform birds, e.g., the domestic fowl, the turkey, and the guinea fowl, the PR are separated by a band of connective tissue, the interparabronchial septa (IPRS) (Figure 10). The IPRS are lacking in the passeriform, the columbiform, the cuculiform, and the psittaciform species; they are nonexistent or very poorly developed in the anseriforms, and; are well-developed in the charadriiform and the galliform species (Maina et al., 1982). The IPRS are lacking in the PR of the lung of the ostrich, *Struthio camelus* (Maina and Nathaniel, 2001) (Figure 11). The PR of the lungs of the flightless and the poor flyers like the penguins and the domestic fowl have wide parabronchial lumen diameters, wide atria, and thin gas exchange mantle relative to the diameters of the PR (Figure 10). In birds on which data are available, the smallest PR with the highest proportion of exchange tissue (ET) occur in the lungs of the psittaciform and passeriform species (Duncker, 1971; Maina et al., 1982). The intraparabronchial arteries and veins are located at the boundaries of the ET mantles of bordering PR (Figure 10). Where IPRS are lacking, the location of the interparabronchial blood vessels demarcate the boundaries of the parabronchi (Figure 11).

Atria, Infundubulae, Air Capillaries, and the Blood-Gas Barrier

A gas ET mantle surrounds the lumina of the PR (Figures 10, 11, 13-15). The thickness of the mantle ranges from 200 to 500 μm (Duncker, 1974, 1979). The atria run outwards from the parabronchial lumen into the ET (Figures 10-16): they are demarcated by interatrial septa (Figure 10, 11, 13-16). In the lungs of small, energetic flyers, the atria are shallow (Duncker, 1974; Maina et al., 1982). The infundibulae are narrower funnel-shaped openings which lead from the atrium into the ET (Figures 14-16): they give rise to air capillaries (AC) (Figures 12, 15, 16) which form part of the ET. Recent three-dimensional computer reconstruction studies have shown that the ACs are rather rotund structures which are interconnected across short and narrow passages (Figures 17-20) while the blood capillaries (BC) are formed by distinctive parts which are rather as long as they are wide (Figures 21-24) (Woodward and Maina, 2005, 2008; Maina and Woodward, 2009).

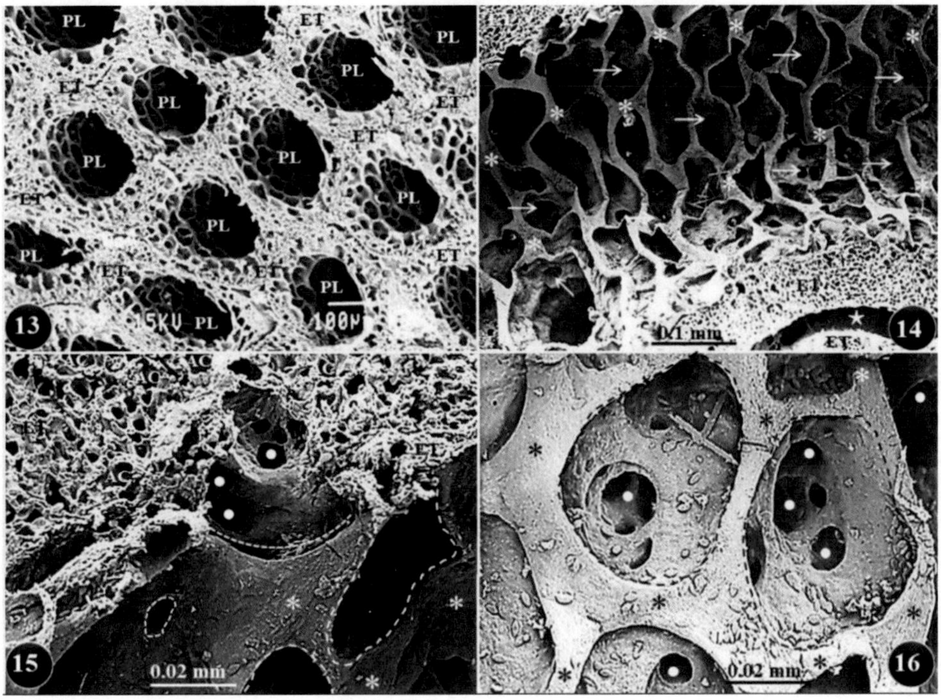

Figures 13-16. Parabronchi of the lung of the domestic fowl. Figure 13. Cross-sections of the parabronchi which comprise of a parabronchial lumen (PL) surrounded by exchange tissue (ET). Figure 14. Longitudinal section of a parabronchus showing the parabrochial lumen giving rise to atria which are surrounded by interatrial septa (asterisks) and which in turn give rise to infundibulae (arrows). ET, exchange tissue; star, interparabronchial septum. Figure 15. Edge of a parabronchial lumen showing atria projecting outwards (dashed areas) which in turn give rise to infundibulae (dots). ET, exchange tissue; asterisks, interatrial septa. Figure 16. Close-up of the surface of the parabronchial lumen showing atria (encircled areas) projecting from it: they give rise to infundibulae (dots). Asterisks, interatrial muscles.

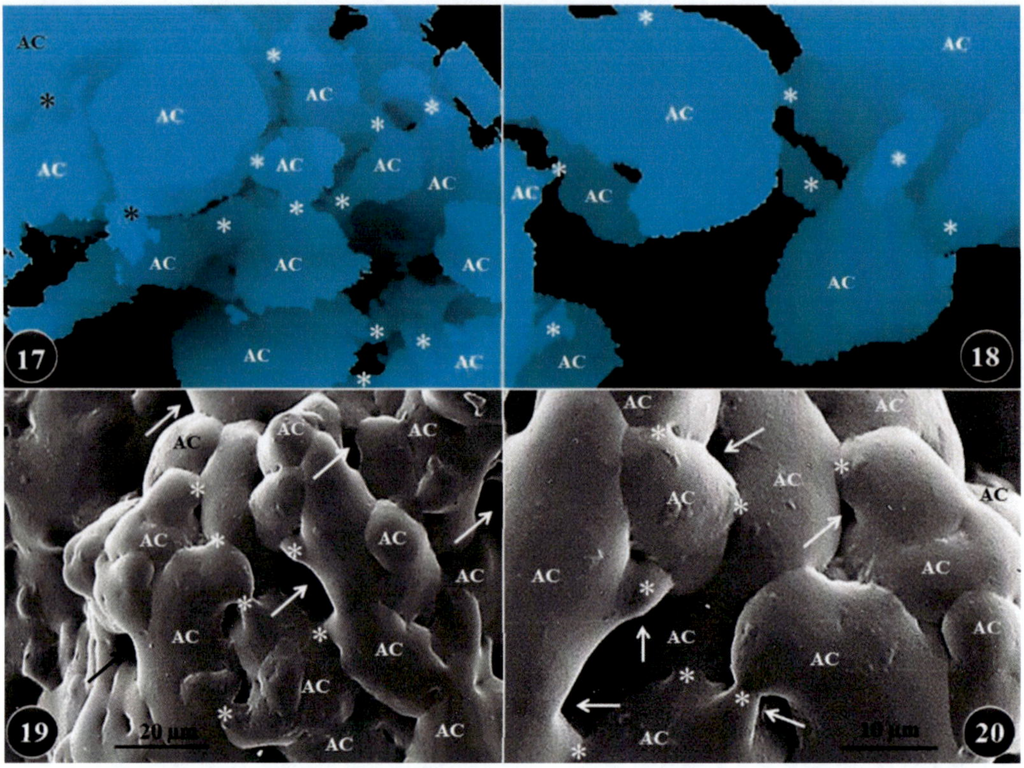

Figures 17-20. Three-dimensional reconstructions (Figures 17 and 18) and latex cast preparations (Figures 19, 20) of the air capillaries (AC) of the lung of the domestic fowl. The respiratory units are globular in shape and are interconnected by narrow passgeways (stars). Arrows, blood capillaries which intertwine witth the ACs.

The diameters of the ACs range from 3 µm in the songbirds to 20 µm in the ostrich, *Struthio camelus* (Duncker, 1972; Maina and Nathaniel, 2001). The BCs and ACs, are the terminal respiratory units of the avian lung. They form a network as they closely intertwine and interdigitate with each other (Figures 21-24). The blood-gas barrier (BGB) separates the air- and the blood capillaries (Figure 24). It comprises of an endothelial cell, a basement membrane, and an epithelial cell (Figures 25, 26). The endothelial cell has abundant micropinocytotic vesicles and exhibits sporadic attenuation while the epithelial cell is very thin, has few organelles, and is covered by the surfactant (Maina and King, 1982). The endothelium, the basement membrane, and the epithelium respectively constitute 67%, 21%, and 12 % of the volume of the BGB (Maina and King, 1982).

Paleopulmo Parabronchi (PPPR) and Neopulmonic Parabronchi (NPPR)

According to Duncker (1971), the avian lung comprises of two anatomically and functionally distinctive regions. Based what he deemed to be their chronological order of evolutionary development, he termed them paleopulmo (ancient = old lung) and neopulmo (modern = new lung). While the paleopulmo occurs in lungs of all birds, the neopulmo is lacking or poorly developed in the basal (ancient = primitive) species.

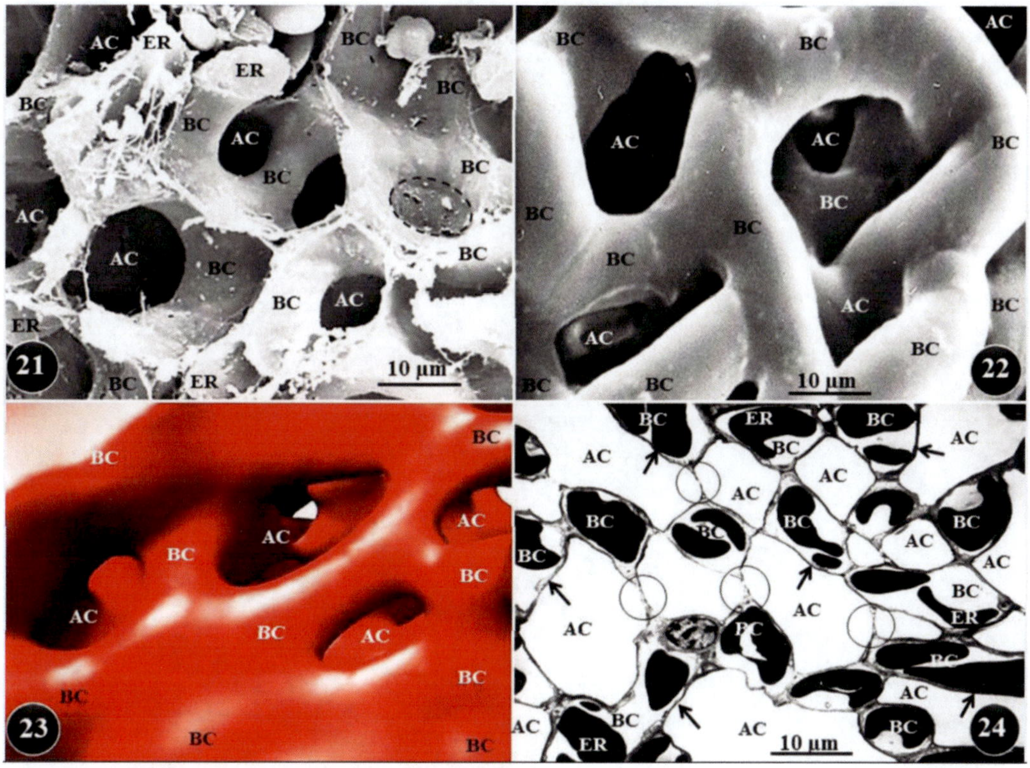

Figures 21-24. Blood capillaries (BC) and air capillaries (AC) in the exchange tissue of the lung of the lung of the domestic fowl. Figure 21 is a critical point preparation. Encircled area, area where air capillaries are separated by epithelial-epithelial cells contacts. Figure 22 is Latex rubber cast preparation; Fig 23: Three-dimensional computer reconstruction, and; Figure 24. A transmission electron micrograpgh. Er, erythrocytes; circles, areas where air capillaries are separated by an epithelial-epithelial cells contacts; arrows, blood-gas barrier.

In birds like stocks (Ciconiidae) and the emu, *Dromaius novaehollandiae*, the NPPR comprise of only a few LDSB which are found in the caudomost part of the PB, their laterally oriented PR, and the connections of the PR to the caudal ASs. In the derived species of birds, e.g., the passerines, the paleopulmo is well-developed that it forms as much as ⅓ of the volume of the lung. There is an inverse correlation between the size of the PB and the development of the NPPR (Duncker, 1971). The proliferation of the NPPR and their extension in a dorsocranial direction displaces on the PB, causing it to bend as it passes through the lung.

In the galliform- and passerine birds, where the NPPR are well-formed, the diameter of the last part of the PB is as narrow as that of a single PR. The complex anastomoses of the NPPR is called 'pulmo reteformis' (Duncker, 1971; King, 1979; López, 1995). In the columbiform-, the galliform, and the passeriform birds, the NPPR and the PPPR interconnect (Figure 7).

While the classification of the PR of the avian lung into the PPPR and the NPPR ones is debatable (López, 1995), it clarified certain aspects of the physiology of the organ. Previously, it could not be explained why the concentration of CO_2 in the abdominal air sacs (AASs) is higher than that in the inspired air. The current explanation is that the inspired air passes through the ET of the NPPR enroute to the AASs, where it collects CO_2.

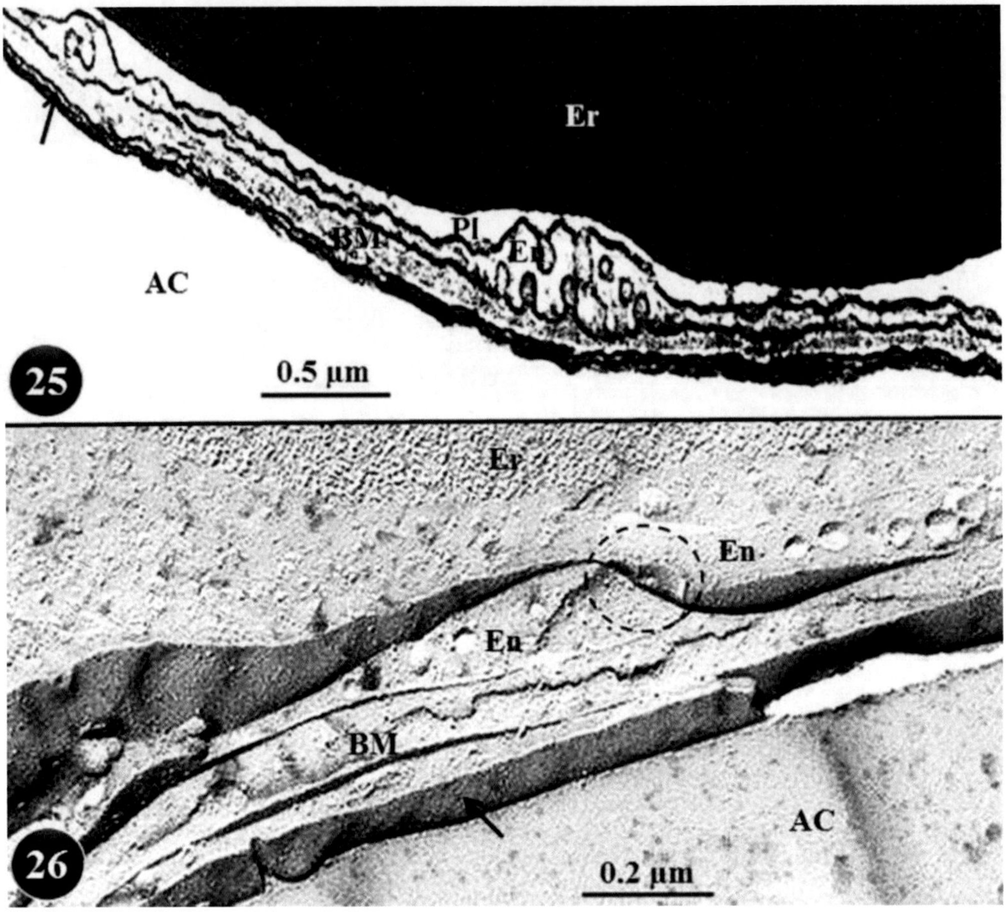

Figures 25-26. The blood-gas barrier of the lung of the domestic fowl. Figure 25: A transmission electron micrograph. Figure 25: A freeze fracture preparation. En, endothelial cell; BM, basement membrane; arrows, epithelium; Er, erythrocyte; AC, air capillary; Pl, plasma layer; dashed circle (Figure 26), intercellular junction.

In birds where the neopulmo is missing or is scanty, the MDSB are located near the surface of the lung: they are easily accessible from the costal surface of the lung for physiological studies, e.g., determination of the air flow and the composition of the respiratory gases.

The structural and functional differences between the PPPR and the NPPR of the avian lung are: a) the PPPR are found on the craniodorsal part of the lung while the NPPR are situated caudoventrally (Figures 5-7), b) the PPPR are arranged as hoop-like stacks which interconnect the MVSB and MDSB (Figures 5-9) while the NPPR anastomose profusely (Figures 5, 6), c) in the PPPR, the air flows continuously and unidirectionally in a caudocranial direction, i.e., from the MDSB to the MVSB whilst in the NPPR it is tidal (= bidirectional): the air flows towards the AASs during inspiration and in the opposite direction, i.e., towards the MDSB, during expiration, and d) the PPPR develop before the NPPR (Romanoff, 1960; Maina 2003a, b). Albeit the morphological differences, the thicknesses of the BGB and the length densities of the BC in the PPPR and NPPR regions of the lung are comparable (Maina, 1982; Maina et al., 1983). Where the PPPR and the NPPR are well-developed, Duncker (1972) suggested that the later may be the exclusive gas exchange area during rest while the PPPR may only be significantly involved during exercise.

Morphometry of the Avian Lung

Birds have lungs which are 27% smaller that those of nonflying mammals of comparable body mass (Maina, 1989; Maina et al., 1989). On average, the gas ET (the parenchyma) of the avian lung comprises ~48% of the volume of the lung (e.g. Maina et al., 1982; Vitali and Richardson, 1998), with the lowest value (18%) in the lung of the emu, *Dromaius novaehollandiae* (Maina and King, 1989) and the highest one (78%) in the lung of the ostrich, *Struthio camelus* (Maina and Nathaniel, 2001). The ACs and BCs constitute between 34% of the ET in the lung of the Penguin, *Spheniscus humboldti* (Maina and King, 1987) to 79% in that of the Emu, *Dromiceius novaehollandiae* (Maina and King 1989). Although the avian lung is small and the parenchymal volume density (fraction) is small, because of the extreme subdivision of the ET, the respiratory surface area (RSA) is 15% greater than that of a lung of a mammal of equivalent body mass. The greatest mass specific RSA of about 90 $cm^2.g^{-1}$ has been been reported in the lung of the Violet-Eared Hummingbird, *Colibri coruscans* (Dubach, 1981) and the African Rock Martin, *Hirundo fuligula* (Maina, 1984, 1989) while the lowest value (5.4 $cm^2.g^{-1}$) occurs in the lung of the Emu (Maina and King, 1989). The thinnest (harmonic mean) BGB of 0.090 μm has been reported in the lungs of the Violet-Eared Hummingbird and the African Rock Martin (Dubach, 1981; Maina, 1984; 1989) while the thickest BGB occurs in the Ostrich lung (0.56 μm) (Maina and Nathaniel, 2001) and that of the penguin lungs (0.53 μm) (Maina and King, 1987). The total volume of blood in the avian lung comprises as much as 36% of the volume of the lung (VL), with 58% to 80% of it found in the BCs (Duncker and Güntert, 1985a, b; Maina et al., 1989). In the African Rock Martin, 29% of the VL comprises of blood, with 79% of it in the BCs (Maina, 1984). Compared to a nonflying mammal, a bird has a pulmonary capillary blood volume which is 22% greater (Maina et al., 1989).

Air Sacs (ASs)

In the different species of birds, the ASs differ in size and location (King, 1966; Duncker, 1971). Transparent membranous structures, they connect to the lung at sites called ostia (Figures 1, 5-7). In birds like the pigeons, *Columba,* the gannet, *Morus bassanus* and the ostrich, *Struthio camelus*, the ASs exit the coelomic cavity and come to lie under the skin (Müller, 1908; Bezuidenhout et al., 2000).

Generally, the cervical ASs are small and paired and are located on the craniodorsal aspect of the coelomic cavity and extend up the neck, pneumatizing the cervical- and thoracic vertebrae, and the vertebral ribs (Duncker, 1971). In some species, e.g., the loons (Gaviidae) and the grebes (Podicipedidae), the cervical ASs are nonexistent. In most birds, the clavicular ASs are large and unpaired: they occupy the cranioventral part of the coelomic cavity, the base of the neck, and much of the right and left axillary space. The craniothoracic- and the caudothoracic ASs are paired: they are found in the space below the lung and the horizontal septum. Commonly, the craniothoracic ASs are smaller than the caudothoracic ones. In the turkey, *Meleagris gallopavo*, the caudothoracic ASs are absent (King and Atherton, 1970). The abdominal ASs occupy the intestinal part of the coelomic cavity, lying on top of the viscera: they are poorly developed in the penguins and the rhea (Duncker, 1971) and are very small in the hummingbirds (Trochilidae) (Stanislaus, 1937). The abdominal ASs considerably pneumatize the synsacrum, the pelvic girdle, the femur, and the caudal vertebrae (King, 1979).

The ASs are lined by a simple epithelium which is supported by a thin layer of connective tissue (Walsh and McLelland, 1974): close to the ostia, the epithelium consists of thin (squamous) cells with ciliated cuboidal- and columnar cells (Fletcher, 1980). A wide band of pseudostratified ciliated columnar epithelium with goblet cells extends from the PB into the AASs in the chicken (Cook and King, 1970).

In the penguins, the epithelial lining of the ASs is generally high, more-or-less cuboidal. Scattered muscle cells and clusters of adipose cells were reported in the walls of the ASs of some species of birds (Fletcher, 1980). The smooth muscle tissue in the wall of the ASs is an extension of the layer which surrounds the parabronchial lumen (Trampel and Fletcher, 1980). The ASs are essentially avascular (Fletcher, 1980). They are not involved in gas exchange (Magnussen et al., 1976). Adrenergic and cholinergic nerve plexes exist in the walls of the ASs (Rawal, 1976; Cook et al., 1987).

Functional Design of the Respiratory System

Among the invaginated gas exchangers of the air-breathing vertebrates, the mechanism by which the avian lung is ventilated is exceptional. Besides being ventilated tidally, i.e., with air moving in-and-out of the respiratory system, as occurs in the mammalian lung, the ET is also ventilated continuously and unidirectionally. The ventilatory mechanism is made possible by the complex morphology of the repiratory system which is separated into a gas exchanging part (the lung) and the ventilatory part (the ASs). Furthermore, the PPPR are arranged in parallel to the PB (Figure 7).

Functionally, the ASs are divided into a cranial group which comprises of the cervical-, the clavicular, and the craniothoracic ASs and a caudal group which consists of the caudothoracic- and the abdominal ASs (Figure 1, 2).

The arrangement of the airways in the avian lung is so complicated (Figures 5-7) that it is impossible to predict the path taken by the inspired air by mere examination of the organ. It takes two inspiratory- and two expiratory cycles for inspired air to pass through the lung-air sac system (Figures 27-30).

The sequence of the air flow is as follows: a) on first inspiration, the air flows to the caudal air sacs (Figure 27); b) air is pushed into the lung with first exhalation (Figure 28); c) on second inhalation, air moves to the cranial air sacs (Figure 29), and; d) on second exhalation, air is expelled to the outside (Figure 30). The lung, specifically the PPPR, which contain most of the ET, is ventilated back-to-front, a through-flow process, which is effected by synchronized bellows-like action of the cranial- and the caudal air groups of ASs (e.g. Fedde, 1980).

In the avian lung, the inspired air flows through the IPPB into the caudal ASs, completely bypassing the orifices of the secondary bronchi, particularly the large ones of the MVSB.

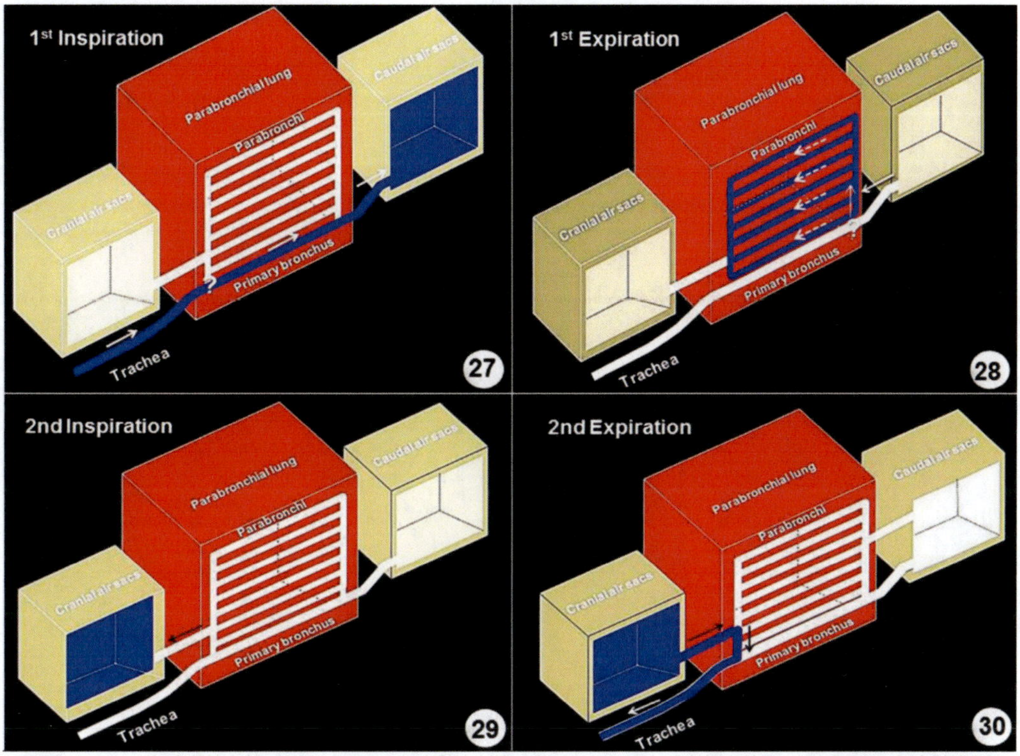

Figures 27-30. Path followed by inhaled air (coloured blue) in the avian respiratory system. On first inspiration, air moves into the caudal (posterior) air sacs (Figure 27); on first expiration it moves into the lung (Figure 28); on second inspiration, air in the lung moves into the cranial (anterior) air sacs (Figure 29), and; on second expiration, the air is expelled (Figure 30). ?, sites where inspiratory- (Figure 27) and expiratory (Figure 28) aerodynamic valving occur; dashed arrows (Figure 28) show the continuous unidirectional airflow in the paleopulmonic parabronchi.

Early investigators like Dotterweich (1934) and Vos (1934) had suggested that anatomical sphincters (valves) controlled the flow of air in the lung. Such structures have, however, never been found in the avian lung even in those areas where they would be expected to occur.

More recent studies have shown that the inspired air is shunted past the openings of the MVSB by a mechanism termed 'inspiratory aerodynamic valving (IAV)' (Banzett et al., 1987, 1991; Wang et al., 1988). Because unidirectional air flow also occurs in pump-ventilated dead avian lungs, structure-specific properties should cause IAV. These include sizes and geometries of the EPPB, the IPPB, and the MVSB (Banzett et al., 1987, 1991; Butler et al., 1988; Kuethe, 1988; Wang et al., 1988, 1992; Maina and Africa, 2000; Maina and Nathaniel, 2001; Maina et al., 2009). Using a radio opaque gas, Wang et al. (1992) showed that narrowing of the EPPB occurred close to the origin of the first MVSB of the lung of the goose, *Anser anser*: they suggested that the narrowing, which they termed 'segmentum accerelans', increases the convective movement of the inspired air, thrusting it across the openings of the MVSB. Wang et al. (1992) observed that the narrowing changed in diameter with the respiratory rate: during ventilatory hyperpnia, the passage was wider and during resting breathing it was narrower. A swelling which bulged into the lumen of the EPPB close to the origin of the first MVSB was reported by Maina and Africa (2001) (Figures 31, 32).

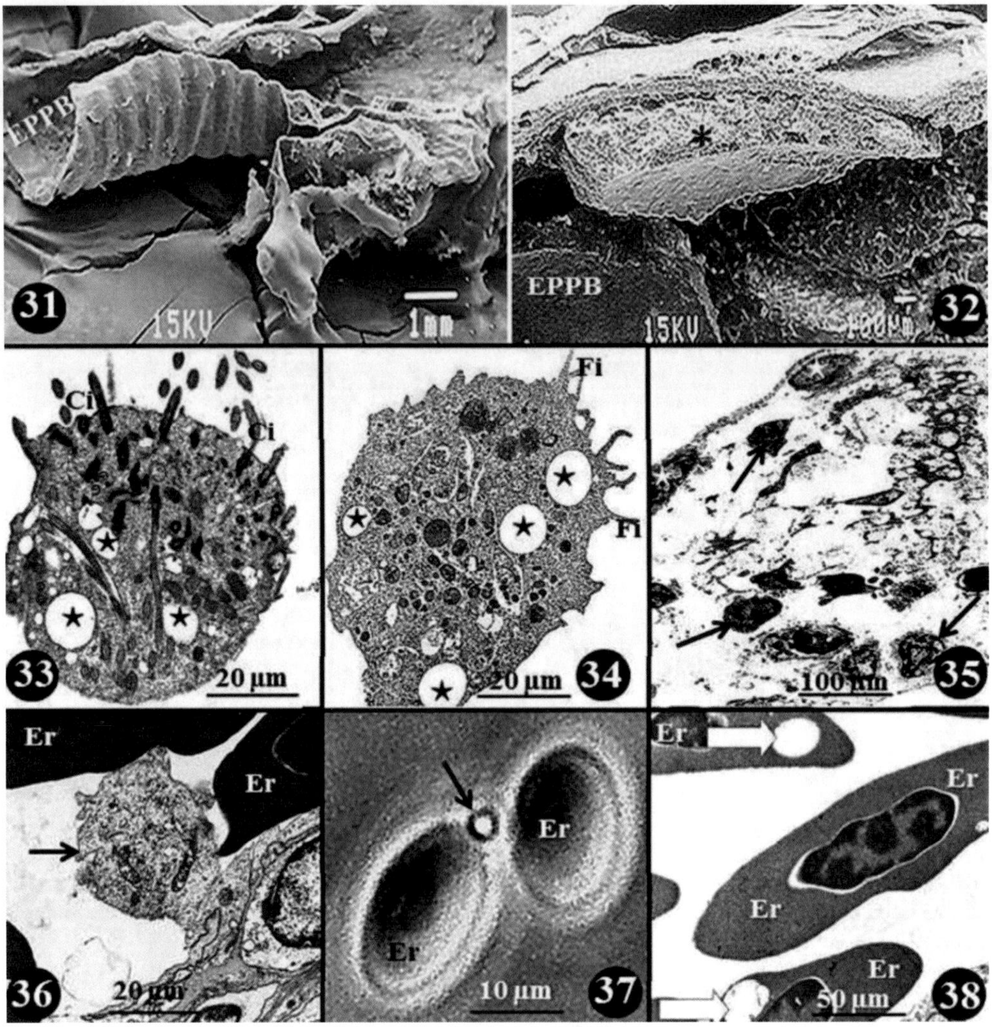

Figures 31-38. Extrapulmonary primary bronchus (EPPB) of the domestic fowl showing location of the segmentum accerelans (asterisk) (Figures 31, 32). Figure 33: A bronchial epithelial cell which has experimentally phagocytized polystylene particles (stars). Ci, cilia. Figure 34: A free (surface) macrophage which has experimentally phagocytized polystylene particles (stars). Fi, flopodia. Figure 35: Epithelial lining of the walls of an infundibulum showing epithelial cells (stars) under which subepithelial macrophages (arrows) exist. Figure 36: An intravascular macrophage (arrow) in a pulmonary blood vessel (arrow). Er, erythrocytes. Figure 37: Erythrocytes (Er) about to ingest a polystylene particle (arrow). Figure 38: Erythrocytes which have phagocytized a polysylene particles (arrows). Figures 33-35, 37, 38 are from the lung of the domestic fowl while Figure 36 is from that of a pigeon.

Cellular Defenses and Susceptibility of the Avian Lung to Infections and Afflictions

With their global distribution and with some species living in close proximity to humans, e.g. the house sparrow, *Passer domesticus*, makes birds potentially ideal animal models for study of environmental pollution (Newman and Schreiber, 1984; Furness and Greenwood

1993). In the poultry industry, huge economical losses have been attributed to respiratory diseases (Mensah and Brain, 1982; Toth et al., 1988). Investigators like Klika et al. (1996) and Spira (1996) remarked that birds are greatly suscreptible to pulmonary diseases and afflictions. Without specific empirical evidence, this was attributed to paucity of free macrophages (FMs) on the respiratory surface of the lung (Stearns et al., 1986; Toth and Siegel, 1986; Klika et al., 1996; Maina and Cowley, 1998; Nganpiep and Maina, 2002). In the lungs of the mouse, the rat, and the guinea pig, $1.1.10^6$, $3.7.10^6$, and $1.4.10^7$ alveolar macrophages were counted by Holt (1979) and Toth and Siegel (1986): the values are ~20 times greater than those on the respiratory systems of birds. In the domestic fowl and the duck, comparatively, Nganpiep and Maina (2002) observed that the body mass normalized number of macrophages in the rat lung significantly exceeded those in the lungs of birds (Figure 39) and considerable flux of macrophages onto the respiratory surface occurred, a response which didn't happen in the rat lung (Figure 40). The mechanism by which macrophages quickly translocated onto the respiratory surface was taken to explain the paucity of FMs on the respiratory surface of the avian lung (Nganpiep and Maina, 2002).

A variety of defense cells which include the bronchial epithelial cells (Figure 33), the FMs (Figure 34), the subepithelial (interstitial) macrophages (Figure 35), the pulmonary intravascular macrophages (Figure 36), and even the red blood cells (Figs 37, 38) (Lippmann and Schlesinger, 1984; Maina and Cowley, 1998; Nganpiep and Maina, 2002) exist in the avian lung. In the domestic fowl, the phagocytized particles are trapped in the mucus lining and then translocated to the epithelial cells (Stearns et al., 1987). The length of the birds trachea is ~3 times longer than that of a mammal of similar body mass (Hinds and Calder, 1971). Albeit paucity of FMs, the avian respiratory system appears to have adequate cellular defense capacity.

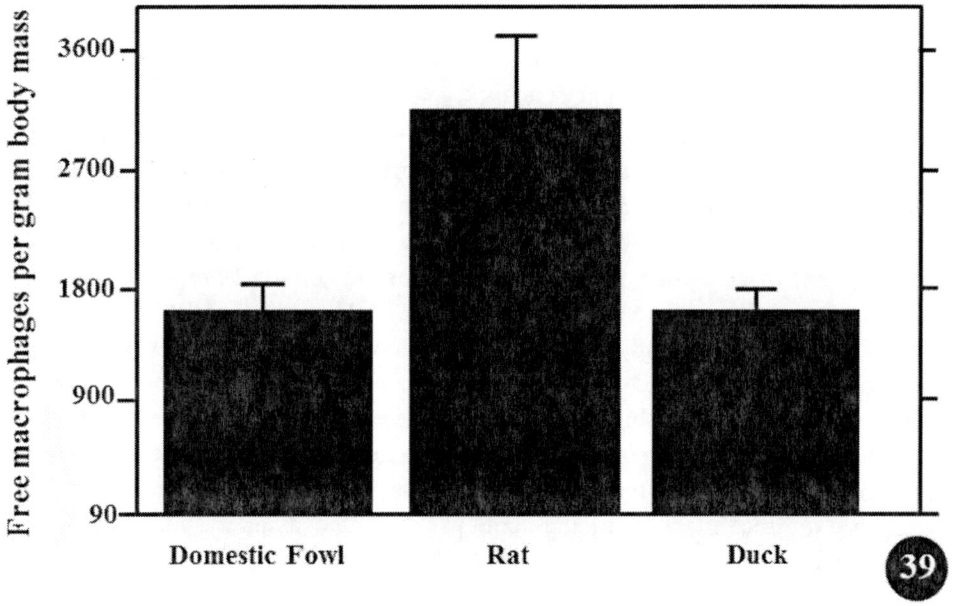

Figure 39. Comparison of numbers of free (surface) macrophages per unit body mass in the lungs of the domestic fowl, the rat, and the duck.

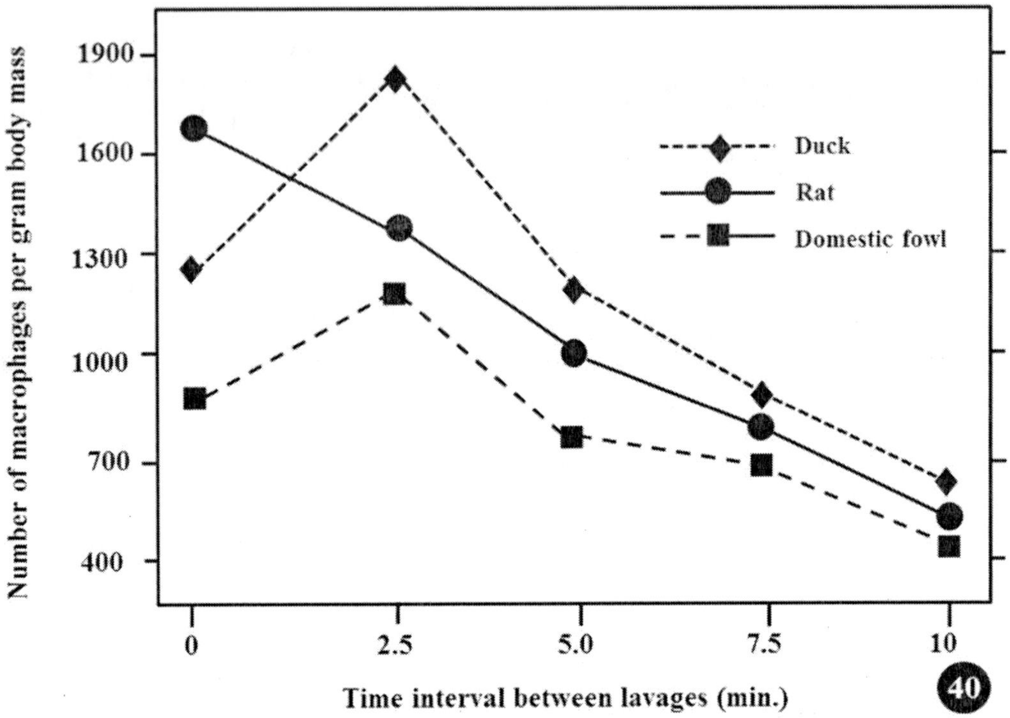

Figure 40. Flux of macrophages to the surface of the lungs in the domestic fowl, the rat, and the duck after repeated lavage. In the domestic fowl and the duck, there was a spike in the number of macrophages on second lavage, which occurred 2.5 minutes after starting the process.

Bacteria are eliminated within 24 to 48 hours (Nagaraja et al., 1984) and conscious chickens clear radioactive technetium particles from the lungs to the intestines within 1 hour of exposure to them (Mensah and Brain, 1982). Chickens with high number of FMs show no signs of respiratory disease (Toth et al., 1988). Some of the morphological- and physiological features which predispose the avian repiratory system to infection by pathogens and injury by inhaled particulates are: a) birds have BGBs which are 56 to 67% thinner and a respiratory surface area which is 15% greater than that of a mammal of comparable body mass (Maina, 1989; Maina et al., 1989), features which make it easier for a pathogen to access and pass from air to blood, b) stemming from their capacious ASs, birds have large mass-specific minute ventilation (tidal volumes) (Brown et al., 1997) and the PPPR are ventilated continuously and unidirectionally (Scheid, 1979; Fedde, 1980): likelihood of entry and deposition of harmful particles and pathogenic microorganisms on the surface of the respiratory system is therefore high, c) in some species of birds, e.g., the ostrich, *Struthio camelus*, where the ASs spread out of the coelomic to lie under the skin (Bezuidenhout et al., 2000): trauma and infection of the skin can lead to air sacculitis (infection of the ASs) and possible spread of the infection to the lung itself. In the avian respiratory system, the predominant sites of infections and deposition of particulates are the caudal ASs (Fedde, 1998). In contrast to the air in the cranial ASs which is filtered as air passes through the PR, the inhaled air which reaches the caudal ASs is relatively more polluted. Furthermore, the inhaled air stays in the caudal ASs for a relatively long time which allows particulates to settle down (Brown et al., 1997). Contrary to the remark that the ASs lack FMs (Brown et al.,

1997), in the domestic fowl, Nganpiep and Maina (2002) showed that they exist in all the air sacs.

Acknowledgment

The colleagues and students who have been involved in my researches on aspects presented here are: M.A. Abdalla, S.G. Kiama, J.D. Woodward, L. Nganpiep, M. Africa, and C. Nathaniel. The writing of this chapter was funded by the National Research Foundation of South Africa.

References

Banzett R. B., Nations C. S., Barnas J. L., Lehr J. L., Jones J. H. (1987). Inspiratory aerodynamic valving in goose lungs depends on gas density and velocity. *Respir. Physiol.*70: 287-300.

Banzett R. B., Nations C. S., Wang N., Fredberg J. J., Butler P. J. (1991). Pressure profiles show features essential to aerodynamic valving in geese. *Respir. Physiol.* 84: 295-309.

Bezuidenhout A. J., Groenewald H. B., Soley J. T. (2000). An anatomical study of the respiratory air sacs in ostriches. *Onderst. J. Vet. Res.* 66: 317-325.

Bramwell C. D. (1971). Aerodynamics of Pteranodon. *J. Linn. Soc., Biol.* 3: 313-328.

Brown R. E., Brain J. D., Wang N. (1997). The avian respiratory system: a unique model for studies of the respiratory toxicosis and for monitoring air quality. *Environ. Hlth. Perspect.* 105: 188-200.

Butler J. P., Banzett R. B., Fredberg J. J. (1988). Inspiratory valving in avian bronchi: aerodynamic considerations. *Respir. Physiol.* 73: 241-256.

Cook R. D., King A. S. (1970). Observations on the ultrastructure of the smooth muscle and its innervation in the avian lung. *J. Anat.* 106: 273-283.

Cook R. D., Vaillant C. R., King A. S. (1987). The structure and innervation of the saccopleural membrane of the domestic fowl, Gallus gallus: an ultrastructural and immunohistochemical study. *J. Anat.* 150: 1-9.

De Beer, G. (1954). Archeopteryx lithographica. British Museum of Natural History, London.

Del Corral J. P. D. (1995). Anatomy and histology of the lung and air sacs of birds. In: LM Pastor (ed.) Histology, Ultrastructure and Immunohistochemistry of the Respiratory Organs in Non-mammalian Vertebrtaes. Publicaciones de la Universitatd de University of Murcia, Murcia (Spain), pp. 179-233.

Dotterweich H. (1930). Versuch überden Weg der Atemluft in der Vogellunge. *Zeitsch. Vergl. Physiol.* 11: 271-284.

Dubach M. (1981). Quantitative analysis of the respiratory system of the house sparrow, budgerigar, and violet-eared hummingbird. *Respir. Physiol.* 46: 43-60.

Duncker H. R. (1971). The lung-air sac system of birds. A contribution to the functional anatomy of the respiratory apparatus. *Ergeb. Anat. Entwicklung* 45: 1-171.

Duncker H. R. (1972). Structure of the avian lung. *Respir. Physiol.* 14: 4-63.

Duncker H. R. (1974). Structure of the avian respiratory tract. *Respir. Physiol.* 22: 1-34.

Duncker H. R. (1979). Die funktionelle Anatomie des Lungen-Luftsack-Systems der Vögel - mit besonderer Berücksichtigung der Greivögel. *Der. Prakt. Tier.* 60: 209-218.

Duncker H. R., Guntert M. (1985a). The quantitative design of the avian respiratory system: from hummingbird to the mute swan. In: Nachtigall W (ed.) BIONA Report No. 3. Gustav-Fischer Verlag, Stuttgart, pp. 361-378.

Duncker H. R., Guntert M. (1985b). Morphometric analysis of the avian respiratory system. In: Duncker H. R., G. Fleischer G. (eds.) Vertebrate. Morphology. Gustav-Fischer, Stuttgart, pp. 383-387.

Fedde M. R. (1980). The structure and gas flow pattern in the avian lung. *Poult. Sci.* 59: 2642-2653.

Fedde M. R. (1998). Relationship of structure and function of the avian respiratory system to disease susceptibility. *Poultry Sci.* 77: 1130-1138.

Fletcher O. J. (1980). Pathology of the avian respiratory system. *Poultry Sci.* 59: 2666-2679.

Furness R. W., Greenwood J. J. D. (eds) (1993). Birds as Monitors of Environmental Change. Chapman and Hall, London.

Gruson E. S. (1976). Checklist of Birds of the World. William Collins, London.

Hinds D. S., Calder W. A. (1971). Tracheal dead space in the respiration of birds. *Evolution* 25: 429-440.

Holt P. G. (1979). Alveolar macrophage. I. A simple technique for the preparation of high numbers of viable alveolar macrophages from small laboratory animals. *J. Immun. Methods* 27: 189-198.

King A. S. (1966) Structural and functional aspects of the avian lung and its air sacs. *Intern. Rev. Gen. Exp. Zool.* 2: 171-267.

King A. S. (1979). Systema respiratorium. In: Baumel J. J., King A. S., Lucas A. M., Breazile J. E., Evans H. E. (eds.) Nomina Anatomica Avium. Academic Press, London, pp. 227-265.

King A. S. and Molovy V. (1971). The anatomy of respiration. In: Bell D. F., Freeman B. M. (eds) Physiology and Biochemistry of the Domestic Fowl, Vol 1. London and New York, Academic Press, pp. 347-384.

King A. S., Atherton J. D. (1970). The identity of the air sacs of the turkey (Melleagris gallopavo). *Acta Anat.* 77: 78-91.

King A. S., King D. Z. (1979). Avian morphology: general principles. In: King A. S., McLelland J (ed.) Form and Function in Birds, Vol I. London-New York, Academic Press, pp. 1-38.

Klika E., Scheuermann D. W., de Groodt-Lasseel M. H. A., Bazantova I., Switka A. (1996). Pulmonary macrophages in birds (barn owl, Tyto tyto alba), domestic fowl (Gallus gallus domestica), quail (Coturnix coturnix), and pigeons (Columba livia). *Anat. Rec.* 256: 87-97.

Kuethe D. O. (1988). Fluid mechanical valving of airflow in bird lungs. *J. Exp. Biol.* 136: 1-12.

Lasiewski R. C. (1962). The energetics of migrating hummingbirds. *Condor* 64: 324.

Lippmann M., Schlesinger R. B. (1984). Interspecies comparison of particle deposition and mucociliary clearance in tracheobronchial airways. *J. Toxicol. Environ. Hlth.* 13: 441-469.

López J. (1995). Anatomy and histology of the lung and air sacs of birds. In: LM Pastor (ed) Histology, Ultrastructure, and Immunohistochemistry of the Respiratory Organs in Nonmammalian Vertebrates. Publicaciones de la Universitatd de University of Murcia, Murcia (Spain), pp. 179-233.

Magnussen H., Willmer H., Scheid P. (1976). Gas exchange in the air sacs: contribution to respiratory gas exchange in ducks. *Respir. Physiol.* 26: 129-146.

Maina J. N. (1982). Stereological analysis of the paleopulmo and neopulmo respiratory regions of the avian lung (Streptopelia decaocto). *IRCS Med. Sci.* 10: 328.

Maina J. N. (1984). Morphometrics of the avian lung. 3. The structural design of the passerine lung. *Respir. Physiol.* 55: 291-309.

Maina J. N. (1989). The morphometry of the avian lung. In: King AS, McLelland J (eds) Form and Function in Birds, Vol.4. Academic Press, London, pp. 307-368.

Maina J. N. (2003a). A systematic study of the development of the airway (bronchial) system of the avian lung from days 3 to 26 of embryogenesis: a transmission electron microscopic study on the domestic fowl, Gallus gallus variant domesticus. *Tissue Cell* 35: 375-391.

Maina J. N. (2005). The Lung-Air Sac System of Birds: Development, Structure, and Function. Springer-Verlag, Heidelberg. Maina JN (2006). Development, structure and function of a novel respiratory organ, the lung-air sac system of birds: To go where no other vertebrate has gone. *Biol. Rev.* 81: 545-579.

Maina J. N. (2008). Functional morphology of the avian respiratory system, the lung-air sac system: efficiency built on complexity. *Ostrich* 79: 117-132.

Maina J. N., Abdalla M. A., King A. S. (1982). Light microscopic morphometry of the lungs of 19 avian species. *Acta Anat.* 112: 264-270.

Maina J. N., Africa M. (2000). Inspiratory aerodynamic valving in the avian lung: functional morphological study of the extrapulmonary primary bronchus. *J. Exp. Biol.* 203: 2865-2876.

Maina J. N., Cowley H. M. (1998). Ultrastructural characterization of the pulmonary cellular defenses in the lung of a bird, the rock dove, Columba livia. *Proc. R. Soc, Lond. B* 265: 1567-1572.

Maina J. N., Howard C. V., Scales L. (1983). Length densities and maximum diameter distribution of the air capillaries of the paleopulmo and neopulmo region of the avian lung. *Acta Stereologica* 2: 101-107.

Maina J. N., King A. S. (1982). The thickness of the avian blood-gas barrier: qualitative and quantitative observations. *J. Anat.* 134: 553-562.

Maina J. N., King A. S. (1987). A morphometric study of the lung of a humboldt penguin (Spheniscus humboldti). *Zentralb. Vet. Med. C Anat. Histo. Embryol.* 16: 293-297.

Maina J. N., King A. S. (1989). The lung of the emu, Dromaius novaehollandiae: a microscopic and morphometric study. *J. Anat.* 163: 67-74.

Maina J. N., King A. S., Settle G. (1989) An allometric study of the pulmonary morphometric parameters in birds, with mammalian comparison. *Philos. Trans. R. Soc. Lond. B* 326: 1-57.

Maina J. N., Nathaniel C. (2001). A qualitative and quantitative study of the lung of an ostrich, Struthio camelus. *J. Exp. Biol.* 204: 2313-2330.

Maina J. N., Singh P., Moss E. A. (2009). Inspiratory aerodynamic valving occurs in the ostrich, Struthio camelus lung: Computational fluid dynamics study under resting unsteady state inhalation. *Respir. Physiol. Neurobiol.* 169: 262-270.

Maina J. N., Woodward J. D. (2009). Three-dimensional serial section computer reconstruction of the arrangement of the structural components of the parabronchus of the ostrich, Struthio camelus lung. *Anat. Rec.* 292: 1685-1698.

Maina, J. N. (2003b). Developmental dynamics of the bronchial (airway)- and air sac systems of the avian respiratory system from days 3 to 26 of life: a scanning electron microscopic study of the domestic fowl, Gallus gallus variant domesticus. *Anat. Embryol.* 207: 119-134.

Marshall A. J. (1962). In: Parker J. T., Haswell P. W. (ed.) A Textbook of Zoology, Vol 2. Macmillan, London.

Mensah G. A., Brain J. D. (1982). Deposition and clearance of inhaled aerosol in the respiratory tract of chickens. *J. Appl. Physiol.* 53: 1423-1428.

Müller B. (1908). The air sacs of the pigeon. Smithson. *Misc. Colls.* 50: 365-414.

Nagaraja K. V., Emery D. A., Jordan A., Sivanandan J. A., Newman J. A., Pomeroy B. S. (1984). Effect of ammonia on the quantitative clearance of Escherichia coli from the lungs, air sacs, and livers of turkeys aerosol vaccinated against Echerichia coli. *Am. J. Vet. Res.* 45: 392-395.

Newman J. R., Schreiber R. K. (1984). Animals as indicators of ecosystem responses to air emissions. *Environ. Manage.* 8: 309-324.

Nganpiep L., Maina J. N. (2002). Composite cellular defense stratagem in the avian respiratory system: functional morphology of the free (surface) macrophages and specialized pulmonary epithelia. *J. Anat.* 200: 499-516.

Ostrom J. H. (1975). The origin of birds. *Annu. Rev. Earth Planetary Sci.* 3: 55-77.

Rawal U. M. (1976). Nerves in the avian air sacs. *Pavo* 14: 57-60.

Romanoff, A. L. (1960). The Avian Embryo. Macmillan, New York.

Scheid P. (1979). Mechanisms of gas exchange in bird lungs. *Rev. Physiol. Biochem. Pharmacol.* 86: 137-186.

Spira A. (1996). Disorders of the respiratory system. In: Rosskopf W., Woerpel R. (eds.) Diseases of Cage and Aviary Birds. Lea and Febiger, Baltimore, pp. 415-428.

Stanislaus M. (1937). Untersuchungen an der Kolibrilunge. *Zeits. Morphol. Tiere* 33: 261-289.

Stearns R. C., Barnas G. M., Walski M., Brain J. D. (1986). Phagocytosis in the gas exchange region of avian lungs. *Fed. Proc.* 45: 959.

Stearns R. C., Barnas G. M., Walski M., Brain J. D. (1987). Deposition and phagocytosis of inhaled particles in the gas exchange region of the duck, Anas platyrhynchos. *Respir. Physiol.* 67: 23-36.

Tobalske B. W., Hedrik T. L., Dial K. P., Biewener A. A. (2003). Comparative power curves in bird flight. *Nature* 421: 363-366.

Toth T. E., Pyle R. H., Caceci T., Siegel P. B., Ochs D. (1988). Cellular defense of the avian respiratory system: influx and nonopsonic phagocytosis by respiratory phagocytes activated by Pasteurella multocida. *Infect. Immun.* 56: 1171-1179.

Toth T. E., Siegel P. B. (1986). Cellular defense for the avian respiratory tract: paucity of free-residing macrophages in the normal chicken. *Avian Dis.* 30: 67-75.

Trampel D. W. and Fletcher O. J. (1980). Ring-stabilizing technique for collection of avian air sacs. *Am. J. Vet. Res.* 14: 1730-1734.

Tucker V. (1970). Energetic cost of locomotion in mammals. *Comp. Biochem. Physiol.* 34: 841-846.

Vitali S. D., Richardson K. C. (1998). Evaluation of pulmonary volumetric morphometry at the light and electron microscopy level in several species of passerine birds. *J. Anat.* 193: 573-580.

Vos H. J. (1934). Über die wege der Atemluft in der Entenlunge. *Zeisch. Vergl. Physiol.* 21: 552-578.

Walsh C., McLelland J. (1974). The ultrastructure of the avian extrapulmonary respiratory epithelium. *Acta Anat.* 89: 412-422.

Wang N., Banzett R. B., Butler J. P., Fredberg J. J. (1988). Bird lung models show that convective inertia effects inspiratory aerodynamic valving. *Respir. Physiol.* 73: 111-124.

Wang N., Banzett R. B., Nations C. S., Jenkins E. A. (1992). An aerodynamic valve in the avian primary bronchus. *J. Exp. Biol.* 262: 441-445.

Welty J. C. (1979). The Life of Birds, 2nd edition. Saunders, Philadelphia.

Wigglesworth V. B. (1972). The Principles of Insect Physiology, 7th Edn. Chapman and Hall, London.

Woodward J. D., Maina J. N. (2005). A three-dimensional digital reconstruction of the components of the gas exchange tissue of the lung of the Muscovy duck, Cairina moschata. *J. Anat* .206: 477-492.

Woodward J. D., Maina J. N. (2008). Study of the structure of the air- and blood capillaries of the gas exchange tissue of the avian lung by serial section three-dimensional reconstruction. *J. Microsc.* 230: 84-93.

Yalden D. W., Morris P. A. (1975). The Lves of Bats. The New York Times Book Co., New York.

Yapp W. B. (1970). The Life and Organization of Birds. Edward Arnold, London.

In: Birds
Editors: Lucas Ruiz and Franco Iglesias
ISBN: 978-1-62948-104-3

Chapter 3

Distribution and Dispersion of Coccidia in Wild Passerines of the Americas

Bruno Pereira Berto[1,*] and Carlos Wilson Gomes Lopes[2,†]

[1]Departamento de Biologia Animal, Instituto de Biologia, Universidade Federal Rural do Rio de Janeiro, Seropédica, RJ, Brazil

[2]Departamento de Parasitologia Animal, Instituto de Veterinária, Universidade Federal Rural do Rio de Janeiro, Seropédica, RJ, Brazil

Abstract

Reports of coccidia of the genera *Eimeria* Schneider, 1875 and *Isospora* Schneider, 1881 from passerine birds have been reported for more than two centuries. In the Americas, the order Passeriformes comprises 47 families containing 2,453 species. *Isospora* species have been observed in 23 (49%) of the 47 families (Dendrocolaptidae, Furnariidae, Thamnophilidae, Cotingidae, Tyrannidae, Corvidae, Meliphagidae, Mimidae, Cardinalidae, Coerebidae, Emberizidae, Estrildidae, Fringillidae, Hirundinidae, Icteridae, Parulidae, Passeridae, Sturnidae, Thraupidae, Timaliidae, Turdidae, Troglodytidae, Zosteropidae) and in 66 (3%) of the 2,453 species. A total of six *Isospora* species have been described from *Ramphocelus bresilius dorsalis* (Sclater); four from *Camarhynchus parvulus* (Gould); and *Saltator similis* (Lafresnaye d'Orbigny); three from each of the following: *Sporophila caerulescens* (Vieillot); *Sporophila frontalis* (Verreaux); *Oryzoborus angolensis* L.; *Thraupis sayaca* L.; and *Zosterops japonicus* (Temminck, Schlegel); two from each of the following: *Dendrocolaptes certhia* (Sclater & Salvin); *Xiphorhynchus ocellatus* (Spix); *Saltator grossus* L.; *Coereba flaveola* L.; *Geospiza fortis* (Gould); *Lonchura punctulata* L.; *Serinus canaria* L.; *Thraupis palmarum* (Wied); *Turdus albicollis* (Vieillot); and *Sicalis flaveola* L.; and a single species from each of the following: *Glyphorynchus spirurus* (Vieillot); *Dendrocincla merula merula* (Lichtenstein); *Dendrocincla merula barletti* (Chubb); *Hyloctistes subulatus* (Spix); *Sclerurus caudacutus* (Vieillot); *Automolus infuscatus* (Sclater); *Hylophylax naevioides* (Lafresnaye); *Procnias nudicollis* (Vieillot); *Myiarchus ferox*

* E-mail: bertobp@ufrrj.br.
† E-mail: lopescwg@ufrrj.br.

(Gmelin); *Mionectes rufiventris* (Cabanis); *Corvus brachyrhynchos* (Brehm); *Calocitta formosa* (Swainson); *Cyanocorax chrysops* (Vieillot); *Foulehaio carunculatus* (Gmelin); *Cardinalis cardinalis* L.; *Paroaria coronata* (Miller); *Tiaris fuliginosus* (Wied); *Zonotrichia capensis* (Müller); *Gubernatrix cristata* (Vieillot); *Carduelis carduelis* L.; *Carduelis atrata* (Lafresnaye d'Orbigny); *Carduelis tristis* L.; *Carduelis chloris* L.; *Fringilla coelebs* L.; *Hemignathus virens* (Cabanis); *Petrochelidon pyrrhonota* (Vieillot); *Dives atroviolaceus* (Lafresnaye d'Orbigny); *Cacicus cela* L.; *Sturnella bellicosa* (Filippi); *Icterus icterus* L.; *Icterus graceannae* (Cassin); *Geothlypis aequinoctialis* (Gmelin); *Passer domesticus* L.; *Gracula religiosa* L.; *Leucopsar rothschildi* (Stresemann); *Chlorospingus ophthalmicus* (Du Bus Gisignies); *Iridosornis analis* (Tschudi); *Dacnis cayana* L.; *Leiothrix lutea* (Scopoli); *Myadestes obscurus* (Gmelin); *Catharus aurantiirostris* (Hartlaub); *Turdus migratorius* L.; *Turdus grayi* (Bonapart); *Turdus assimilis* (Cabanis); *Turdus plebejus* (Cabanis); *Thryophilus rufalbus* (Lafresnaye); *Cantorchilus modestus* (Cabanis); and *Mimus gilvus* (Vieillot). Observations in relation to *Eimeria* species have been significantly less frequent. Specifically, the species *Eimeria divinolimai* Berto, Flausino, Ferreira, Lopes, 2008 and *Eimeria sicki* Berto, Luz, Flausino, Ferreira, Lopes, 2009 were described from two hosts only [*Casiornis rufus* (Vieillot) and *Myiarchus ferox* (Gmelin)] both of which belong to the family Tyrannidae. This chapter contributes to our understanding of the distribuition of coccidian parasites of passerine birds from Northern, Central and Southern America. To this end, the coccidia were organized and grouped according to the family of the host, emphasizing the geographic location of the reports and associated aspects.

Introduction

The order Passeriformes is the most representative of the class Aves. It includes more than 5,000 species distributed throughout the world, which account for more than 50% of the class Aves. The passerine birds of the Americas are mostly endemic and occupy, in the case of South America, a large number of ecological niches that in other continents are inhabited by other birds. At present a total of 1,020 species are found in Brazil, of which 234 are endemic. A small number of families have numerical predominace, with the families Tyrannidae, Formicariidae and Furnariidae (Sick, 1997; IUCN, 2013), demonstrating the greatest levels of species diversity. Similarly to other vertebrates the passerines can be infected by coccidia, primarily by members of the genus *Isospora* Schneider, 1881 and, to a lesser extent, by the genus *Eimeria* Schneider, 1875.

The specificity of these coccidia to a taxon of Passeriformes is controversial. Prior to 1982, more than 100 species of passerines had been reported as hosts for *I. lacazei*, disregarding host specificity. In recent times, the descriptions of coccidia have generally been made in accordance with the guidelines proposed by Duszynski & Wilber (1997). Those authors put forward the concept of intra-family specificity, when they suggested that a new coccidian species should be compared in detail to the coccidian species that is most structurally similar to it within the same host family. Berto et al. (2010a; 2011b) provided support for the family-specific concept when they reported three passerines of same family, but distinct genera, as hosts to the same coccidian species (Berto et al., 2011a).

In general, coccidiosis is an important disease that manifests with symptoms characterized by diarrhea, dehydration, and death particularly in younger animals (Soulsby, 1987). Evaluation of disease in wild birds is difficult. Despite frequent descriptions of

Isospora species parasitizing these birds, these are generally not observed clinical signs (Massey, 2003). Swayne et al. (1991) associated the death of 13 warblers *Vermivora ruficapilla* Wilson, 1811 with coccidiosis. Gill and Paperna (2008) associated the presence of oocysts of *Isospora* with diarrhea and death of sparrows *P. domesticus* in Israel. Working with house sparrows *Passer domesticus* L. in Romania, Pap et al. (2013) demonstrate that besides these well-known consequences, parasites might also have serious, long-lasting effects through influencing feather quality and, ultimately, fitness of the host.

Currently, our knowledge of the reports of coccidia of passerines in the Americas is widely distributed throughout an extensive body of literature. The aims of this chapter was contributes to our understanding of the distribuition and dispersion of coccidian parasites of passerine birds from Northern, Central and Southern America. To this end, the coccidia were organized and grouped according to the family of the host, emphasizing the geographic location of the reports and associated aspects.

Eimeria Schneider, 1875

Eimeria parasitizing Passeriformes in the Americas have been reported just recently. Berto et al. (2008c, 2009d) described *Eimeria divinolimai* Berto, Flausino, Ferreira, Lopes, 2008 and *Eimeria sicki* Berto, Luz, Flausino, Ferreira, Lopes, 2009 parasitizing the rufous casiornis *Casiornis rufus* (Vieillot) and the short-crested flycatcher *Myiarchus ferox* (Gmelin), respectively. Both species of birds belong to the family Tyrannidae and inhabit Southeastern Brazil.

Isospora Schneider, 1881

Hundreds of species of *Isospora* have been described in passerine birds; however, the majority of those descriptions occurred in Eurasia. Twenty-three families of Passeriformes are hosts for 86 isosporoid coccidia in the Americas: (1) Dendrocolaptidae, (2) Furnariidae and (3) Thamnophilidae of parvorder Furnariida, infraorder Tyranni; (4) Cotingidae and (5) Tyrannidae of parvorder Tyrannida, infraorder Tyranni; (6) Corvidae and (7) Meliphagidae of parvorder Corvida, infraorder Passeri; (8) Cardinalidae, (9) Coerebidae, (10) Emberizidae, (11) Estrildidae, (12) Fringillidae, (13) Hirundinidae, (14) Icteridae, (15) Mimidae, (16) Parulidae, (17) Passeridae, (18) Sturnidae, (19) Thraupidae, (20) Timaliidae, (21) Turdidae, (22) Troglodytidae and (23) Zosteropidae of parvorder Passerida, infraorder Passeri.

The sections which follow provide reports of *Isospora* according to the family of their passerine hosts, emphasizing the geographic location of the reports and associated aspects.

Dendrocolaptidae Gray, 1840

The coccidia parasites of this family were initially reported in birds from Ecuador, where McQuistion & Capparella (1995) recovered oocysts from the feces of two Amazonian barred woodcreeper *Dendrocolaptes certhia* (Sclater & Salvin). The observed species were named *Isospora concentrica* McQuistion, Capparella, 1995 and *Isospora magna* McQuistion, Capparella, 1995. Two years later, the species *Isospora ocellati* McQuistion, Walden, Capparella, 1997 and *Isospora striata* McQuistion, Walden, Capparella, 1997 were described parasitizing the ocellated woodcreeper *Xiphorhynchus ocellatus* (Spix) in Ecuador (McQuistion et al., 1997). In the same year a fifth species, *Isospora ubique* McQuistion, Capparella, 1997 was described as a parasite of wedge-billed woodcreepers *Glyphorynchus spirurus* (Vieillot) in both, Ecuador and Republic of Guyana (McQuistion & Capparella, 1997). More recently, McQuistion et al. (2010) described the species *Isospora dendrocinclae* McQuistion, Galewsky, Capparella, Rebling, 2010 from the fecal contents of the white-chinned woodcreeper, *Dendrocincla merula merula* (Lichtenstein) collected in Guyana and *Dendrocincla merula barletti* (Chubb) from Peru.

Furnariidae Gray, 1840

Three species of *Isospora* were described in this family: (1) *Isospora hyloctistum* McQuistion, Capparella, 1994 from the striped woodhaunter *Hyloctistes subulatus* (Spix); (2) *Isospora scleruri* McQuistion, Capparella, 1994 from the black-tailed leaftosser *Sclerurus caudacutus* (Vieillot); and (3) *Isospora automoli* McQuistion, Barber, Capparella, 1999 from the olive-backed foliage-gleaner *Automolus infuscatus* (Sclater). These birds were also from Ecuador (McQuistion & Capparella, 1994; McQuistion et al., 1999).

Thamnophilidae Swainson, 1824

A single *Isospora* species has been described in this family: *Isospora sagittulae* McQuistion, Capparella, 1992 was found in *Hylophylax naevioides* (Lafresnaye), once again in Ecuador (McQuistion & Capparella, 1992a).

Cotingidae Bonaparte, 1849

Doležalová et al. (2004) recovered oocysts from feces of the bare-throated bellbird *Procnias nudicollis* (Vieillot), in Spain. These birds had been recently imported from Brazil by The Barcelona City Zoo and were examined because they were in quarantine. The parasite was named *Isospora araponga* Doležalová, Torres, Fernández, Modrý, 2004.

Tyrannidae Vigors, 1825

The coccidian parasites of this family were recently described by Berto et al. (2009d, 2009e). Specifically, *Isospora feroxis* Berto, Luz, Flausino, Ferreira, Lopes, 2009 was found in the short-crested flycatcher *M. ferox*, while *Isospora mionectesi* Berto, Flausino, Luz, Ferreira, Lopes, 2009 was described from the grey-hooded flycatcher *Mionectes rufiventris* (Cabanis). Both birds inhabit Southeastern Brazil.

Corvidae Leach, 1820

Wobester & Cawthorn (1985) described *Isospora brachyrhynchi* Wobester, Cawthorn, 1985 from fecal samples of American crows *Corvus brachyrhynchos* (Brehm) collected in the province of Saskatchewan, Western Canada.

In South America two species have been described from this family: The first was *Isospora cyanocoracis* Upton, Current, Clubb, 1985 , parasitizing *Cyanocorax chrysops* (Vieillot), in Argentina (Upton et al. 1985) and the second species was *Isospora calocitta* Upton, Wright, Langen, 1995 , described as a parasite of *Calocitta formosa* (Swainson), in Costa Rica (Upton et al. 1995a).

Meliphagidae Vigors, 1825

Adamczyk et al. (2004) described the only species of coccidian parasite of this family: *Isospora samoaensis* Adamczyk, McQuistion, LaPointe, 2004 was described parasitizing *Foulehaio carunculatus* (Gmelin), in American Samoa, an island located in Micronesia, South Pacific Ocean.

Cardinalidae Ridgway, 1901

Isospora vanriperorum (Levine, Van Riper, Van Riper, 1980) Levine, 1982 was the first coccidium described from cardinalid birds (Levine, 1982b). This species was first observed in the northern cardinal *Cardinalis cardinalis* L. in Hawaii. Originally, Levine et al. (1980) named it *Isospora cardinalis*; however this name had previously been ascribed to the coccidium parasite of the extinct passerine *Lophospingus pusillus*, described in the German Republic by Gottschalk (1972),

Lopes et al. (2007) recovered oocysts of *I. vanriperorum* from the green-winged saltator *Saltator similis* (Lafresnaye d'Orbigny) in Rio de Janeiro, Brazil. This transmission was sustained due to the introduction of northern cardinals into South America for breeding in captivity (Carvalho, 2009).

Two other species have been described in cardinalid birds from Ecuador: (1) *Isospora pityli* McQuistion, Capparella, 1992; and (2) *Isospora formarum* McQuistion, Capparella, 1992. Both species were described as parasites of the slate-coloured grosbeak *Saltator grossus* L. (McQuistion & Capparella, 1992b). Subsequently, Berto et al. (2008d) described a further two species in the feces of *S. similis* in Brazil: (1) *Isospora saltatori* Berto, Balthazar, Flausino, Lopes, 2008; and (2) *Isospora trincaferri* Berto, Balthazar, Flausino, Lopes, 2008.

Recently, Coelho et al. (2013) describe another coccidian species of the same host, which was named *Isospora similisi* Coelho, Berto, Neves, Oliveira, Flausino, Lopes, 2013. These passerines were apprehended from illegal trading and kept in quarantine in CETAS (Centro de Triagem de Animais Silvestres - Center for Triage of Wild Animals)/IBAMA (Instituto Brasileiro do Meio Ambiente e dos Recursos Naturais Renováveis - Brazilian Institute of Environment and Natural Resources)/MMA (Ministério do Meio Ambiente - Ministry of Environment) located in the Municipality of Seropédica, state of Rio de Janeiro.

Coerebidae Lafresnaye and d'Orbigny, 1838

The species, *Isospora cagasebi* Berto, Flausino, Luz, Ferreira Lopes, 2008 and *Isospora coerebae* Berto, Flausino, Luz, Ferreira, Lopes, 2010 were described from the bananaquit *Coereba flaveola* L. collected in southeastern Brazil, (Berto et al. 2008b; Berto et al. 2011).

Emberizidae Vigors, 1825

Upton et al. (1985), based in Florida, were the first researchers to describe coccidian parasites of this family when they recovered oocysts of *Isospora paroariae* Upton, Current, Clubb, 1985 from fecal samples of the red-crested cardinal *Paroaria coronata* (Miller), which had been collected in Argentina.

Dr. Thomas E. McQuistion subsequently described several species of *Isospora* from endemic passerine birds of the Galapagos Islands (McQuistion & Wilson, 1988; 1989; McQuistion, 1990). Initially, McQuistion and Wilson (1988) described four new species parasitizing *Camarhynchus parvulus* (Gould): (1) *Isospora rotunda* McQuistion, Wilson,

1988; (2) *Isospora fragmenta* McQuistion, Wilson, 1988; (3) *Isospora exigua* McQuistion, Wilson, 1988; and (4) *Isospora temeraria* McQuistion, Wilson, 1988. One year later, McQuistion & Wilson (1989), described *Isospora geospizae* McQuistion, Wilson, 1989 parasitizing *Geospiza fortis* (Gould). Finally, *Isospora daphnensis* McQuistion, 1990 was also described from *G. fortis*, in the Galapagos Islands (McQuistion, 1990).

Ball and Daszak (1997), based in the United Kingdom, described *Isospora tiaris* Ball, Daszak, 1997 observed in the feces of sooty grassquits *Tiaris fuliginosus* (Wied) from Venezuela.

Carvalho-filho et al. (2005) described three new species parasitizing double-collared seedeaters *Sporophila caerulescens* (Vieillot) maintained at the CETAS/ IBAMA/ MMA, located in the mMunicipality of Seropédica in the State of Rio de Janeiro. These species were named: (1) *Isospora sporophilae* Carvalho-filho, Meireles, Ribeiro, Lopes, 2005; (2) *Isospora flausinoi* Carvalho-filho, Meireles, Ribeiro, Lopes, 2005; and (3) *Isospora teixeirafilhoi* Carvalho-filho, Meireles, Ribeiro, Lopes, 2005.

Isospora curio Silva, Literák, Koudela, 2006; *Isospora braziliensis* Silva, Literák, Koudela, 2006; and *Isospora paranaensis* Silva, Literák, Koudela, 2006 were described from fecal samples of lesser seed-finches *Oryzoborus angolensis* L. bred in captivity, in the state of Mato Grosso do Sul, Brazil (Silva et al., 2006).

Also in Brazil, Berto et al. (2009a) described *Isospora frontalis* Berto, Balthazar, Flausino, Lopes, 2009; *Isospora teresopoliensis* Berto, Balthazar, Flausino, Lopes, 2009; and *Isospora chanchaoi* Berto, Balthazar, Flausino, Lopes, 2009 from buffy-fronted seedeaters *Sporophila frontalis* (Verreaux), bred in captivity, in the Municipality of Teresópolis, located in a mountainous region of the State of Rio de Janeiro. In the same mountainous location, the species *Isospora ticoticoi* Balthazar, Berto, Flausino, Lopes, 2009 was described from captive rufous-collared sparrows *Zonotrichia capensis* (Müller) (Balthazar et al., 2009).

From South Brazil, a new coccidian species was reported from the endangered yellow cardinal *Gubernatrix cristata* (Vieillot) (IUCN, 2011). *Isospora bocamontensis* Pereira, Berto, Flausino, Lovato, Lopes, 2011 was described infecting yellow cardinals at the district of Boca do Monte, Municipality of Santa Maria in the State of Rio Grande do Sul (Pereira et al., 2011).

Recently, Coelho et al. (2011a) describe two new coccidian species from the saffron finch *Sicalis flaveola* L., a very common species in South America. *Isospora cetasiensis* Coelho, Berto, Neves, Oliveira, Flausino, Lopes, 2011 and *Isospora sicalisi* Coelho, Berto, Neves, Oliveira, Flausino, Lopes, 2011 infect saffron finches kept in quarantine in CETAS/ IBAMA/ MMA, located at the Municipality of Seropédica in the State of Rio de Janeiro.

Estrildidae Bonaparte, 1850

Two *Isospora* species were recognized as parasites of estrildid birds in Hawaii: (1) *Isospora ivensae* Levine, Van Riper, Van Riper, 1980; and (2) *Isospora lyonensis* Upton, Marchiondo, Williams, 1988 were both described from the scaly-breasted munia *Lonchura punctulata* L. (Levine et al., 1980; Upton et al. 1988).

Fringillidae Leach, 1820

The overwhelming majority of reports of *Isospora* have been made in this family. *Isospora lacazei* (Labbé, 1893) Levine, 1982 had been reported in over 100 different species of birds (Boughton, 1930; Henry, 1932; Boughton et al., 1938; Rysavy, 1954; Scholtyseck,

1954; Levine & Mohan, 1960; Mandal, 1965; Anwar, 1966a; 1966b; Mandal & Bhattarcharya, 1969; Hernandez-Rodrigues et al., 1976a; 1976b) until Levine (1982a) definitively classified it as a parasite of the European goldfinch *Carduelis carduelis* L., considering the description of Hernandez-Rodriguez et al. (1976a; 1976b). It should be noted, that this description is similar to that of oocysts recorded from *Carduelis chloris* L. and *Fringilla coelebs* L. by Anwar (1966b).

Isospora canaria Box, 1975 and *Isospora serini* (Aragão, 1933) Box, 1975 were described and subsequently re-described during studies of the intra- and extra-intestinal life cycles in the island canary *Serinus canaria* L. in USA (Box, 1966; 1967; 1970; 1975; 1977; 1981). *Isospora canaria* was reported recently from *S. canaria* in a breeder in Southeastern Brazil (Berto et al., 2013).

Levine et al. (1980) described *Isospora loxopis* Levine, Van Riper, Van Riper, 1980 , from *Hemignathus virens* (Cabanis) sampled in Hawaii. In Italy, Rossi et al. (1996) examined finches imported from South America and recovered oocysts from the feces of *Carduelis atrata* (Lafresnaye d'Orbigny), which were nominated *Isospora atrata* Rossi, Macchione, Perrucci, 1996. In the province of Ontario, Canada, *Isospora gryphoni* Olson, Gissing, Barta, Middleton, 1998 was described parasitizing *Carduelis tristis* L. (Olson et al., 1998).

Hirundinidae Rafinesque, 1815

The species *Isospora petrochelidon* Stabler, Kitzmiller, 1972 was described as a parasite of *Petrochelidon pyrrhonota* (Vieillot) by Stabler & Kitzmiller (1972) in the USA.

Icteridae Vigors, 1825

The first description of *Isospora* parasites of this family was made during the 1960's by Pellérdy (1967), in Cuba, where *Isospora divitis* Pellérdy, 1967 was described from *Dives atroviolaceus* (Lafresnaye d'Orbigny).

Thereafter, *Isospora cacici* Lainson, 1994 was described from the yellow-rumped cacique *Cacicus cela* L., in the Brazilian Amazon (Lainson, 1994) and Upton et al. (1995b) described *Isospora bellicosa* Upton, Stamper, Whitaker, 1995 from *Sturnella bellicosa* (Filippi), in the USA.

More recently, two new species were described from icterid birds reared in the Zoo /National Aquarium of Baltimore, in the USA: Firstly, *Isospora icterus* Upton, Whitaker, 2000 was described parasitizing *Icterus icterus* L., while *Isospora graceannae* Upton, Whitaker, 2000 parasitized *Icterus graceannae* (Cassin) (Upton & Whitaker, 2000).

Mimidae Bonaparte, 1853

Coelho et al. (2011b) reported the only *Isospora* of this family, which infects tropical mockingbirds *Mimus gilvus* Vieillot, recovered from illegal trade in Brazil. *Isospora mimusi* Coelho, Berto, Neves, Oliveira, Flausino, Lopes, 2011 was described from feces of mockingbirds held at CETAS/ IBAMA/ MMA, located in the Municipality of Seropédica in the State of Rio de Janeiro, for rehabilitation and reintroduction into the wild.

Parulidae Wetmore, Friedmann, Lincoln, Miller, Peters, Van Rossem, Van Tyne and Zimmer 1947

Berto et al. (2009f) reported the only coccidium of this family: *Isospora piacobrai* Berto, Flausino, Luz, Ferreira, Lopes, 2009 was observed as a parasite of the masked yellowthroat *Geothlypis aequinoctialis* (Gmelin), sampled on Marambaia Island in the State of Rio de Janeiro, Southeastern Brazil.

Passeridae Rafinesque, 1815

Levine (1982a) proposed the name *Isospora passeris* (Levine, Mohan, 1960) Levine, 1982 for the intestinal *Isospora* described from the house sparrow *P. domesticus*. This author used the characters reported by Levine & Mohan (1960) in the USA, resulting in the relocation of several previous descriptions of *I. lacazei*.

Sturnidae Rafinesque, 1815

It is generally accepted that Upton and colleagues, are recognized as the preeminent authorities on the two species of *Isospora* parasites associated with sturnid birds.

Isospora graculai (Bhatia, Chauhan, Arora, Agrawal, 1973) Upton, Erst, Clubb, Current, 1984 was originally described from the hill myna *Gracula religiosa* L., in India (Bhatia et al., 1973). Upton et al. (1984) subsequently re-described this species, providing more features, when they recovered morphologically similar oocysts in the feces of hill mynas imported from Southeast Asia into the USA.

More recently, *Isospora rothschildi* Upton, Wilson, Norton, Greiner, 2001 was described in fecal samples from *Leucopsar rothschildi* (Stresemann), in USA (Upton et al., 2001).

Thraupidae Cabanis, 1847

To date, 12 species of coccidia have been described parasitizing birds of this family. *Isospora thraupis* Lainson, 1994 is considered to represent the first species and was described in tanagers. The sporulated oocysts that provided the basis for this description were obtained from the feces of the palm tanager *Thraupis palmarum* (Wied), in the Amazon (Lainson, 1994).

Templar et al. (2004) described *Isospora andesensis* Templar, McQuistion, Capparella, 2004 from *Chlorospingus ophthalmicus* (Du Bus Gisignies), 1847 and Metzelaars et al. (2005) described *Isospora irisidornisi* Metzelaars, Spaargaren, McQuistion, Capparella, 2005 from *Iridosornis analis* (Tschudi). Both studies were conducted on Peruvian birds.

Subsequently, Berto et al. (2008a, 2009b, 2009c; 2010b) described the presence of nine new species from Brazilian tanagers: (1) *Isospora tiesangui* Berto, Flausino, Luz, Ferreira Lopes, 2008; (2) *Isospora marambaiensis* Berto, Flausino, Luz, Ferreira Lopes, 2008; (3) *Isospora sepetibensis* Berto, Flausino, Luz, Ferreira Lopes, 2008; (4) *Isospora cadimi* Berto, Flausino, Luz, Ferreira Lopes, 2009; (5) *Isospora navarroi* Berto, Flausino, Luz, Ferreira Lopes, 2009 and (6) *Isospora ramphoceli* Berto, Flausino, Luz, Ferreira Lopes, 2010 from the Brazilian tanager *Ramphocelus bresilius dorsalis* (Sclater) on Marambaia Island in the state of Rio de Janeiro, Southeastern Brazil; and (7) *Isospora sanhaci* Berto, Balthazar, Flausino, Lopes, 2009; (8) *Isospora sayacae* Berto, Balthazar, Flausino, Lopes, 2009; and (9) *Isospora silvasouzai* Berto, Balthazar, Flausino, Lopes, 2009 from the sayaca tanager *Thraupis sayaca* L., at the Municipality of Teresópolis in the State of Rio de Janeiro.

Recently, Berto et al. (2010a) reported two new hosts for *I. tiesangui* on Marambaia Island: (1) The palm tanager *T. palmarum*, and the (2) the blue dacnis *Dacnis cayana* L.

Timaliidae Horsfield and Vigors, 1827

Isospora leiothrixi McQuistion, McAllister, Buice, 1996 represents the only described species in this family among passerine birds of the Americas: Its host was *Leiothrix lutea* (Scopoli), which was maintained at the Dallas Zoo, in the USA (McQuistion et al., 1996).

Turdidae Rafinesque, 1815

The first species described in this family was named *Isospora phaeornis* Levine, Van Riper, Van Riper, 1980, and parasitized *Myadestes obscurus* (Gmelin), in Hawaii (Levine et al., 1980). Later, McQuistion & Holmes (1988) described *Isospora robini* McQuistion, Holmes, 1988 as a parasite of the American robin *Turdus migratorius* L., in USA.

Lainson & Shaw (1989) described two species parasitizing the white-necked thrush *Turdus albicollis* (Vieillot), in the sState of Pará, Brazil: *Isospora tucuruiensis* Lainson, Shaw, 1989 and *Isospora albicollis* Lainson, Shaw, 1989.

Recently, Keeler et al. (2012a) described *Isospora zorzali* Keeler, Yabsley, Gibbs, McGraw, Hernandez, 2012 from thrushes of the species *Catharus aurantiirostris* (Hartlaub), *Turdus grayi* (Bonapart), *Turdus assimilis* (Cabanis) and *Turdus plebejus* (Cabanis), in the San Luis Valley, approximately 7 km southwest of the Monteverde region in northwestern Costa Rica.

Troglodytidae Swainson, 1831

Keeler et al. (2012b) reported *Isospora troglodytes* Keeler, Yabsley, Fox, McGraw, Hernandez, 2012 of this family, which infects the wrens *Thryophilus rufalbus* (Lafresnaye) and *Cantorchilus modestus* (Cabanis) in San Luis Valley, Monteverde region in northwestern Costa Rica.

Zosteropidae Bonaparte, 1853

Three species are recognized as parasites of this family. All three parasitized *Zosterops japonicus* (Temminck, Schlegel) in Hawaii: Firstly, Levine et al. (1980) described *Isospora brayi* Levine, Van Riper and Van Riper, 1980; and then Upton et al. (1988) described *Isospora manoaensis* Upton, Marchiondo, Williams, 1988 and *Isospora mejiro* Upton, Marchiondo, Williams, 1988.

Conclusion

As noted in this chapter and illustrated in Figure 1, some coccidia have been described in passerines which inhabit geographically isolated areas and thus, remain isolated; however, the majority of the passerine hosts demonstrate a wide geographical distribution. Figure 1 aims to demonstrate the location that *Isospora* were reported from passerines, also informing ones which were reported in more than one country. However, this Figure in no way suggests that countries without reports do not have *Isospora* parasitizing passerines, just should not have been investigated at the present moment.

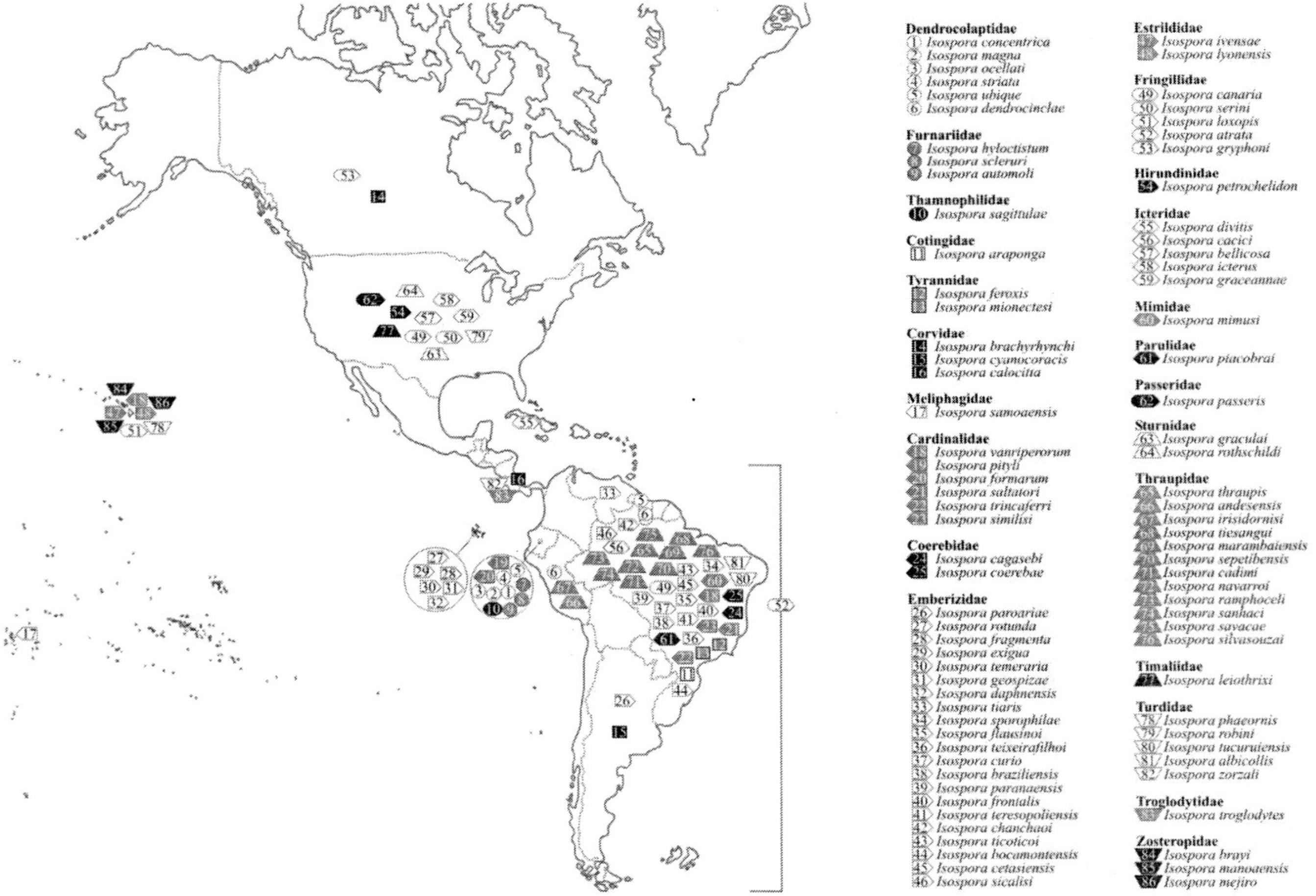

Figure 1. Countries where the *Isospora* species have been described or reported. Numbers with same symbol denote coccidia of the same host family. *Isospora atrata* (52) has been described from finches imported from South America.

Figure 2. Geographical range of thraupid hosts for *Isospora* in South America.

The coccidia reported from birds of the Americas are highly relevant, given that transmission of parasites can occur between sympatric birds of the same family (Carvalho-Filho et al., 2005; Berto et al., 2010c; 2011b). As an example, Figure 2 demonstrates the geographical range of thraupid hosts for *Isospora* in South America. It was observed that the wide range of *T. sayaca*, *T. palmarum* and *D. cayana* ensure transmission of *Isospora* of Thraupidae in Brazil and several other countries in South and Central America.

Oppositely, transmission between non-sympatric species that inhabit distant continents or islands is unlikely, as occurs with coccidia reported in Hawaii and American Samoa (Figure 1) (Levine et al., 1980; Duszynski & Wilber, 1997; Duszynski et al., 1999; Carvalho-Filho et al., 2005; Berto et al., 2010c; 2011b).

In other hand, the trade of exotic birds should enhance the transmission of its coccidian among species of the same family. The presence of these exotic coccidia in the Americas is relevant to native passerines since transmission could occur from breeders in wild areas. Berto et al. (2013) report *Isospora canaria* Box, 1975 from canaries *Serinus canaria* L. from a breeder in Southeastern Brazil. In this work, was emphasized the risk of transmission to two native fringillid passerines, *Carduelis magellanica* (Vieillot) and *Carduelis yarrellii* (Audubon), which are native in South America, including Brazil. The presence of *C. yarrellii* is more relevant, since this species is listed as vulnerable (IUCN, 2013). Moreover, a hybrid species (called 'pintagol' in Brazil) from the crossing in captivity of male *C. magellanica* with female *S. canaria* should enhance the possibility of transmission of *I. canaria* to native passerines. Reinforcing this hypothesis, probably *I. vanriperorum* was transmitted from northern cardinals *C. cardinalis* to green-winged saltators *S. similis* through breeders in South America, since *C. cardinalis* and *S. similis* are not sympatric in the wild (Lopes et al., 2007).

The trafficking of wild animals (biopiracy) is the third largest illegal trade in the world, second only to drug trafficking and weapons. In Brazil, about 12 million animals are trafficked every year and the specimens of Passeriformes and Psittaciformes are the most commom seized by the environmental authorities. In this context, the centers screening of wild animals have the purpose of receiving, triage and treat the wild animals rescued or seized by the inspectors, and eventually receiving wild animals that were captive domestic irregularly shaped like animals pet. The work of receiving and triage animals should entail in the record identifying each individual species and sex, including as much information about location of capture and time captivity. The specimens should be housed and stimulated, according to the habitat of each species. After this, the animals should be quarantined getting proper nutrition and being observed and examined to identify potential parasites and diseases (Ferreira & Glock, 2004; Araújo et al., 2010).

This last point is crucial because it guides the destiny of this specimen for reintroduction into the wild or in zoos. The failure to identify a parasite of an animal seized, followed by their release in the wild, different from its original, would provide the introduction of a new parasite to susceptible hosts. Anyway, even if an animal is reintroduced parasitized in their habitat of origin is important that they are treated in quarantine to reduce the parasite load, since the stress of captivity reduces host immunity, increasing the number of parasites that may impact the wild (Coelho et al., 2012a; b). In this view, the work of the centers screening of wild animals become more important since they can provide the dispersion of parasites, including coccidia when the birds were reintroduced in the wild.

Through routine of quarantine performed at centers screening of wild animals, *I. araponga* of *P. nudicollis* was described in Spain (Doležalová et al., 2004) and *I. similisi* of *S.*

similis, *I. sporophilae*, *I. flausinoi*, *I. teixeirafilhoi* of *S. caerulescens* (Carvalho-filho et al., 2005), *I. cetasiensis*, *I. sicalisi* of *S. flaveola* (Coelho et al., 2011a), and *I. mimusi* of *Mimus gilvus* (Coelho et al., 2011b) were described in Brazil.

Finally, this chapter does not aimed to provide scientific basis for the identification of the coccidia; however, details on the phenotypic characteristics of sporulated oocysts of coccidia of passerines of the Americas, which allowed the identification of *Eimeria* and *Isospora* species, can be acquired in the original articles of species descriptions (Berto et al., 2011a).

References

Adamczyk, K. J., McQuistion, T. E., & LaPointe, D. (2004). A new coccidian parasite, *Isospora samoaensis*, from the wattled honeyeater (*Foulehaio carunculata*) from American Samoa. *Acta Protozoologica, 43,* 179-181.

Anwar, M. (1966a). Cytochemical studies on infections with *Isospora lacazei*. *Transactions of the Royal Society of Tropical Medicine and Hygiene, 60,* 428-429.

Anwar, M. (1966b). *Isospora lacazei* (Labbe, 1893) and *I. chloridis* sp. n. (Protozoa: Eimeriidae) from the English Sparrow (*Passer domesticus*), Greenfinch (*Chloris chloris*) and Chaffinch (*Fringilla coelebs*). *Journal of Protozoology, 13,* 84-90.

Araujo, A. C. B., Behr, E. R., Longhi, S. J., Kanieski, M. R., & Menezes, P. (2010). Diagnóstico sobre a avifauna apreendida e entregue espontaneamente na Região Central do Rio Grande do Sul, Brasil. *Revista Brasileira de Biociências, 8,* 279-284.

Ball, S. J., & Daszak, P. (1997). *Isospora tiaris* n. sp. (Apicomplexa: Eimeriidae) from the sooty grassquit (*Tiaris fuliginosa*), a passeriform bird of South America. *Journal of Parasitology, 83,* 465-466.

Balthazar, L. M. C., Berto, B. P., Flausino, W., & Lopes, C. W. G. (2009). *Isospora ticoticoi* n. sp. (Apicomplexa: Eimeriidae) from the rufous-collared sparrow *Zonotrichia capensis* in South America. *Acta Protozoologica, 48,* 345-349.

Berto, B. P., Flausino, W., Luz, H. R., Ferreira, I., & Lopes, C. W. G. (2008a). Three New Coccidian Parasites of Brazilian Tanager (*Ramphocelus bresilius dorsalis*) from South America. *Acta Protozoologica, 47,* 77-81.

Berto, B. P., Flausino, W., Luz, H. R., Ferreira, I., & Lopes, C. W. G. (2008b). *Isospora cagasebi* sp. nov. (Apicomplexa, Eimeriidae) from the bananaquit, *Coereba flaveola* of Brazil. *Acta Parasitologica, 53,* 117-119.

Berto, B. P., Flausino, W., Ferreira, I., & Lopes, C. W. G. (2008c). *Eimeria divinolimai* sp. n. (Apicomplexa: Eimeriidae) in the *Rufous casiornis Casiornis rufus* Vieillot, 1816 (Passeriformes: Tyrannidae) in Brazil. *Revista Brasileira de Parasitologia Veterinária, 17,* 33-35.

Berto, B. P., Balthazar, L. M. C., Flausino, W., & Lopes, C. W. G. (2008d). Two New Coccidian Parasites of Green-Winged Saltator (*Saltator similis*) from South America. *Acta Protozoologica, 47,* 263-267.

Berto, B. P., Balthazar, L. M. C., Flausino, W., & Lopes, C. W. G. (2009a). Three new species of *Isospora* Schneider, 1881 (Apicomplexa: Eimeriidae) from the buffy-fronted seedeater *Sporophila frontalis* Verreaux, 1869 (Passeriformes: Emberizidae) from South America. *Systematic Parasitology, 73,* 65-69.

Berto, B. P., Flausino, W., Luz, H. R., Ferreira, I., & Lopes, C. W. G. (2009b). Two new *Isospora* species from Brazilian tanager (*Ramphocelus bresilius dorsalis*) of South America. *Parasitology Research, 105,* 635-639.

Berto, B. P., Balthazar, L. M. C., Flausino, W., & Lopes, C. W. G. (2009c). New isosporoid coccidian parasites of sayaca tanager, *Thraupis sayaca*, from South America. *Acta Parasitologica, 54,* 90-94.

Berto, B. P., Luz, H. R., Flausino, W., Ferreira, I., & Lopes, C. W. G. (2009d). New species of *Eimeria* Schneider, 1875 and *Isospora* Schneider, 1881 (Apicomplexa: Eimeriidae) from the short-crested flycatcher *Myiarchus ferox* (Gmelin) (Passeriformes: Tyrannidae) in South America. *Systematic Parasitology, 74,* 75-80.

Berto, B. P., Flausino, W., Luz, H. R., Ferreira, I., & Lopes, C. W. G. (2009e). *Isospora mionectesi* sp. Nov. (Apicomplexa, Eimeriidae) from the grey-hooded flycatcher, *Mionectes rufiventris* in Brazil. *Acta Parasitologica, 54,* 301-304.

Berto, B. P., Luz, H. R., Flausino, W., Ferreira, I., & Lopes, C. W. G. (2009f). *Isospora piacobrai* n. sp. (Apicomplexa: Eimeriidae) from the masked yellowthroat *Geothlypis aequinoctialis* (Gmelin) (Passeriformes: Parulidae) in South America. *Systematic Parasitology, 75*, 225-230.

Berto, B. P., Luz, H. R., Ferreira, I., Flausino, W., & Lopes, C. W. G. (2010a). Two new hosts for *Isospora tiesangui* Berto, Flausino, Luz, Ferreira & Lopes, 2008 (Apicomplexa: Eimeriidae). *Revista Brasileira de Medicina Veterinária, 32,* 168-171.

Berto, B. P., Flausino, W., Luz, H. R., Ferreira, I., & Lopes, C. W. G. (2010b). *Isospora ramphoceli* n. sp. from the Brazilian tanager (*Ramphocelus bresilius dorsalis*) in South America. *Zootaxa, 2650,* 57-62.

Berto, B. P., Luz, H. R., Ferreira, I., Flausino, W., & Lopes, C. W. G. (2010c). A diagnostic tool to identify species of the genus *Isospora* Schneider, 1881 (Apicomplexa: Eimeriidae) based on sporulated oocysts from Thaupidae family (Aves: Passeriformes): a dichotomous key. *Revista Brasileira de Medicina Veterinária, 32,* 182-186.

Berto, B. P., Flausino, W., McIntosh, D., Teixeira-Filho, W. L., & Lopes, C. W. (2011a). Coccidia of New World passerine birds (Aves: Passeriformes): a review of *Eimeria* Schneider, 1875 and *Isospora* Schneider, 1881 (Apicomplexa: Eimeriidae). *Systematic parasitology, 80,* 159-204.

Berto, B. P., Luz, H. R., Flausino, W., Teixeira-Filho, W. L., Ferreira, I., & Lopes, C. W. G. (2011b). Isosporoid coccidia (Apicomplexa: Eimeriidae) parasites of tanagers (Passeriformes: Thraupidae) from the Marambaia Island, Brazil. *Pesquisa Veterinária Brasileira, 31*, 798-805.

Berto, B. P., Flausino, W., Luz, H. R., Ferreira, I., & Lopes, C. W. G. (2011c). *Isospora coerebae* n. sp. (Apicomplexa: Eimeriidae) from the bananaquit *Coereba flaveola* (Passeriformes: Coerebidae) in South America. *Revista Brasileira de Parasitologia Veterinária, 20*, 22-26.

Berto, B. P., Ferreira, I., Flausino, W., Teixeira-Filho, W. L., & Lopes, C. W. (2013). *Isospora canaria* Box, 1975 (Apicomplexa: Eimeriidae) from canaries *Serinus canaria* Linnaeus (Passeriformes: Fringillidae) in Brazil. *Systematic Parasitology, 85,* 49-53.

Bhatia, B. B., Chauhan, P. P. S., Arora, G. S., & Agrawal, R. D. (1973). Species composition of coccidia of some mammals and birds at the Zoological Gardens, Dehli and Lucknow. *Indian Journal of Animal Science, 43,* 944-947.

Boughton, D. C. (1930). The value of measurements in the study of a protozoan parasite *Isospora lacazei* (Labbe). *American Journal of Hygiene, 11,* 212-226.

Boughton, D. C., Boughton, R. B. & Volk, J. (1938). Avian hosts of the genus *Isospora* (Coccidiida). *Ohio Journal of Science, 38,* 149-163.

Box, E. D. (1966). Blood and tissue protozoa of the English sparrow (*Passer domesticus domesticus*) in Galveston, Texas. *Journal of Protozoology, 13,* 204-208.

Box, E. D. (1967). Influence of *Isospora* infections on potency of avian *Lankesterella* (Atoxoplasma Garhnam, 1950). *Journal of Parasitology, 53,* 1140-1147.

Box, E. D. (1970). *Atoxoplasma* associated with an isosporan oocyst in canaries. *Journal of Protozoology, 17,* 391-396.

Box, E. D. (1975). Exogenous stages of *Isospora serini* (Aragão) and *Isospora canaria* sp. n. in the canary (*Serinus canarius* Linnaeus). *Journal of Protozoology, 22,* 165-169.

Box, E. D. (1977). Life cycles of two *Isospora* species in the canary, *Serinus canarius* Linnaeus. *Journal of Protozoology, 24,* 57-67.

Box, E. D. (1981). *Isospora* as an extraintestinal parasite of passerine birds. *Journal of Protozoology, 28,* 244-246.

Carvalho, J. S. (2009). Cardeal da Virgínia. http://www.petbrazil.com.br. Cited 22 Apr. 2013.

Carvalho-Filho, P., Meireles, G., Ribeiro, C., & Lopes, C. W. G. (2005). Three new species of *Isospora* Schneider, 1881 (Apicomplexa: Eimeriidae) from the double-collared seedeater, *Sporophila caerulescens* (Passeriformes: Emberizidae), from eastern Brazil. *Memórias do Instituto Oswaldo Cruz*, 100, 151-154.

Coelho, C. D., Berto, B. P., Neves, D. M., de Oliveira, V. M., Flausino, W., & Lopes, C. W. G. (2011a). Two new *Isospora* species from the saffron finch, *Sicalis flaveola* in Brazil. *Acta Parasitologica, 56*, 239-244.

Coelho, C. D., Berto, B. P., Neves, D. M., De Oliveira, V. M., Flausino, W., & Lopes, C. W. (2011b). *Isospora mimusi* n. sp.(Apicomplexa: Eimeriidae) from the Tropical Mockingbird *Mimus gilvus* in South America. *Acta Protozoologica, 50*, 137.

Coelho, C. D., Berto, B. P., Neves, D. M., De Oliveira, V. M., Flausino, W., & Lopes, C. W. (2012a). Diagnóstico e tratamento das coccidioses em trinca-ferros-verdadeiros *Saltator similis* D orbigny Lafresnaye, 1837 mantidos em regime de quarentena. *Revista Brasileira de Medicina Veterinária, 34,* 46-54.

Coelho, C. D., Berto, B. P., Neves, D. M., De Oliveira, V. M., Flausino, W., & Lopes, C. W. (2012b). Tratamento da coccidiose causada por espécies do gênero *Isospora* Schneider, 1881 em curiós *Sporophila angolensis* Linnaeus, 1766 e bicudos *Sporophila maximiliani* Cabanis, 1851 mantidos em regime de quarentena. *Revista Brasileira de Medicina Veterinária, 34,* 102-108.

Coelho, C. D., Berto, B. P., Neves, D. M., Oliveira, V. M. D., Flausino, W., & Lopes, C. W. G. (2013). Oocyst shedding by green-winged-saltator (*Saltator similis*) in the diagnostic of coccidiosis and *Isospora similisi* n. sp.(Apicomplexa: Eimeriidae). *Revista Brasileira de Parasitologia Veterinária, 22*, 64-70.

Doležalová, M., Torres, J., Fernández, H., Modrý, D. (2004). *Isospora araponga* sp. n. (Apicomplexa: Eimeriidae), a new species of *Isospora* Schneider from a bare-throated bellbird, *Procnias nudicollis* (Vieillot, 1817) (Passeriformes: Cotingidae) from Brazil. *Memórias do Instituto Oswaldo Cruz, 99,* 829-830.

Duszynski, D. W., Upton, S. J., & Couch, L. (1999). The coccidia of the world. http://www.k-state.edu/parasitology/worldcoccidia/index.html. Cited 22 Apr. 2013.

Duszynski, D. W., & Wilber, P. G. (1997). A guideline for the preparation of species descriptions in the Eimeridae. *Journal of Parasitology, 83,* 333-336.

Ferreira, C. M., & Glock, L. (2006). Diagnóstico preliminar sobre a avifauna traficada no Rio Grande do Sul, Brasil. *Biociências, 12,* 21-30.

Gill, H., & Paperna, I. (2008). Proliferative visceral *Isospora* (atoxoplasmosis) with morbid impact on the Israeli sparrow *Passer domesticus biblicus* Hartert, 1904. *Parasitology Research, 103*, 493-499.

Gottschalk, C. (1972). Beitrag zür Faunistik der Vogelkokzidien Thuringens und Sachsens. *Betrage zür Vogelkunde, 18,* 61-69.

Henry, D. P. (1932). *Isospora buteonis* sp. nov. from the hawk and owl, and notes on *Isospora lacazii* (Labbe) in birds. *University of California Publications in Zoology, 37,* 291-300.

Hernandez-Rodriguez, S., Calero-Carretero, R., Becerra-Martell, C., Dominguez-de-Tena, M., Moreno-Montanez, T., & Martinez-Gomez, F. (1976a). Ultrastructure of the intestinal phases of *Isospora lacazei* Labbe 1893 (Eimeriidae, parasite) in *Carduelis carduelis* (goldfinch). *Revista Iberica de Parasitologia, 36,* 67-79.

Hernandez-Rodriquez, S., Martinez-Gomez, F., Becerra-Martell, C., Calero-Carretero, R., Moreno-Montanez, T., & Dominguez-de-Tena, M. (1976b). *Isospora lacazei* Labbe, 1893 en Passeriformes de la provincia de Cordoba. *Revista Iberica de Parasitologia, 36,* 81-8.

IUCN. (2013). International Union for Conservation of Nature and Natural Resources. http://www.iucnredlist.org. Cited 28 Apr. 2013.

Keeler, S. P., Yabsley, M. J., Gibbs, S. E., McGraw, S. N., & Hernandez, S. M. (2012a). A new *Isospora* species of passerines in the family Turdidae from Costa Rica. *Journal of Parasitology, 98*, 167-169.

Keeler, S. P., Yabsley, M. J., Fox, J. M., McGraw, S. N., & Hernandez, S. M. (2012b). *Isospora troglodytes* n. sp.(Apicomplexa: Eimeriidae), a new coccidian species from wrens of Costa Rica. *Parasitology Research, 110,* 1723-1725.

Lainson, R., & Shaw, J. J. (1989). Two new species of *Eimeria* and three new species of *Isospora* (Apicomplexa, Eimeriidae) from Brazilian mammals and birds. *Bulletin du Museum National D'Histoire Naturalle, 11,* 349-365.

Lainson, R. (1994). Observations on some avian coccidia (Apicomplexa: Eimeriidae) in Amazonian Brazil. *Memórias do Instituto Oswaldo Cruz, 89,* 303-311.

Levine, N. D., & Mohan, R. N. (1960). *Isospora* sp. (Protozoa: Eimeriidae) from cattle and its relationship to *I. lacazei* of the English sparrow. *Journal of Parasitology, 46,* p. 733-741.

Levine, N. D., Van Riper, S., & Van Riper, C. (1980). Five New Species of *Isospora* from Hawaiian Birds. *Journal of Protozoology, 27,* 258-259.

Levine, N. D. (1982a). *Isospora passeris* n. sp. from the house sparrow *Passer domesticus, I. lacazei*, and related Apicomplexan protozoa. *Transactions of the American Microscopical Society, 101,* 66-74.

Levine, N. D. (1982b). *Isospora vanriperorum* n. nom. for *I. cardinalis* Levine, Van Riper & Van Riper, 1980, preoccupied. *Journal of Protozoology, 29,* 653.

Lopes, B. do B., Berto B. P., Massad, F., & Lopes C. W. G. (2007). *Isospora vanriperorum* Levine, 1982 (Apicomplexa: Eimeriidae) in the green-winged saltator, *Saltator similis* (Passeriformes: Cardinalinae) in southeastern Brazil. *Revista Brasileira de Parasitologia Veterinária, 16,* 211-214.

Mandal, A. K. (1965). Studies on some aspects of avian coccidia (Protozoa: Sporozoa). 6. The occurence of *Isospora* sp. in goats with redescription of *Isospora lacazei* (Labbe, 1893) from some common Indian birds. *Journal of the Assam Scientific Society, 8,* 71-75.

Mandal, A. K., & Bhattacharya, A. (1969). A biometrical study of the oocysts of *Isospora lacazei* (Labbé), a common parasite of the house-sparrow, *Passer domesticus* Linnaeus. *Progress in Protozoology*, 3, 341-342.

Massey, J. G. (2003). Diseases and medical management of wild passeriformes. *Seminars in Avian and Exotic Pet Medicine, 12*, 29-36.

McQuistion, T. E., & Holmes, B. B. (1988). *Isospora robini* sp. n., a new coccidian parasite (Apicomplexa: Eimeriidae) from the American robin (*Turdus migratorius*). *Proceedings of the Helminthological Society of Washington, 55,* 324-325.

McQuistion, T. E., & Wilson, M. (1988). Four new species of *Isospora* from the small tree finch (*Camarhynchus parvulus*) from the Galapagos Island. *Journal of Parasitology, 35*, 98-99.

McQuistion, T. E., & Wilson, M. (1989). *Isospora geospizae*, a new coccidian parasite (Apicomplexa: Eimeriidae) from the small ground finch (*Geospiza fuliginosa*) and the medium ground finch (*Geospiza fortis*) from the Galapagos Island. *Systematic Parasitology, 14,* 141-144, 1989.

McQuistion, T. E. (1990). *Isospora daphnensis* n. sp. (Apicomplexa: Eimeriidae) from the medium ground finch (*Geospiza fortis*) from the Galapagos Island. *Journal of Parasitology, 76,* 30-32.

McQuistion, T. E., & Capparella, A. (1992a). *Isospora sagittulae*, a new coccidian parasite (Apicomplexa: Eimeriidae) from the spotted antbird (*Hylophylax naevioid*es). *Transactions of the American Microscopical Society, 111,* 365-368.

McQuistion, T. E., & Capparella, A. (1992b). Two new coccidian parasites from the slate-colored grosbeak (*Pitylus grossus*) of South America. *Journal of Parasitology, 78,* 805-807.

McQuistion, T. E., & Capparella, A. (1994). Two new species of Isospora (Apicomplexa: Eimeriidae) from ovenbirds (Passeriformes: Furnariidae) of South America. *Transactions of the American Microscopical Society, 113,* 90-95.

McQuistion, T. E., & Capparella, A. (1995). Two new isosporan parasites (Apicomplexa: Eimeriidae) from the South American woodcreeper *Dendrocolaptes certhia. Acta Protozoologica, 34,* 299-302.

McQuistion, T. E., McAllister, C. T., & Buice, R. E. (1996). A new species of *Isospora* (Apicomplexa) from captive Pekin robins, *Leiothrix lutea* (Passeriformes: Sylviidae), from the Dallas Zoo. *Acta Protozoologica, 35,* 73-75.

McQuistion, T. E., Walden, K. K. O., & Capparella, A. P. (1997). Two new species of *Isospora* (Apicomplexa: Eimeriidae) from the ocellated woodcreeper, *Xiphorhynchus ocellatus* (Passeriformes: Dendrocolaptidae). *Archiv für Protistenkunde, 147,* 401-404.

McQuistion, T. E., & Capparella, A. (1997). *Isospora ubique*: a new coccidian parasite from the wedge-billed woodcreeper (*Glyphorynchus spirurus*) from South America. *Acta Protozoologica, 36,* 75-78.

McQuistion, T. E., Barber, C. Y., & Capparella, A. P. (1999). *Isospora automoli*, a new coccidian parasite (Apicomplexa: Eimeriidae) from the buff-throated foliage-gleaner *Automolus ochrolaemus* and the olive-backed foliage gleaner *A. infuscatus* from South America. *Systematic Parasitology, 44,* 71-73.

McQuistion, T. E.,Galewsky, S., Capparella, A. P., & Rebling, F. (2010). *Isospora dendrocinclae* n. sp. (Apicomplexa: Eimeriidae) from the Whitechinned Woodcreeper (*Dendrocincla merula*) from South America. *Acta Protozoologica, 49,* 121-124.

Metzelaars, H., Spaargaren, T., McQuistion, T. E., & Capparella, A. P. (2005). *Isospora iridosornisi,* a new coccidian parasite (Apicomplexa, Eimeriidae) from the yellow-throated tanager, *Iridosornis analis* of South America. *Acta Parasitologica, 50,* 191-193.

Olson, V. A., Gissing, G. J., Barta, J. R., & Middleton, A. L. A. (1998). A new *Isospora* sp. from *Carduelis tristis* (Aves: Fringillidae) from Ontario, Canada. *Journal of Parasitology, 84,* 153-156.

Pap, P. L., Vágási, C. I., Bărbos, L., & Marton, A. (2013). Chronic coccidian infestation compromises flight feather quality in house sparrows *Passer domesticus*. *Biological Journal of the Linnean Society, 108*, 414-428.

Pellérdy, L. (1967). Three new coccidia parasitic in Cuban birds (Protozoa: Sporozoa). *Acta Zoologica Academiae Scientiarum Hungaricae, 13,* 227-230.

Pereira, L. Q., Berto, B. P., Flausino, W., Lovato, M. & Lopes, C. W. G. (2011) *Isospora bocamontensis* n. sp. (Apicomplexa: Eimeriidae) from the yellow cardinal *Gubernatrix cristata* (Vieillot) (Passeriformes: Emberizidae) in South America. *Systematic Parasitology, 78,* 73-80.

Rossi, G., Perrucci, S., & Macchioni, G. (1996). *Isospora atrata* (Apicomplexa, Eimeriidae): a new coccidium isolated from *Carduelis atrata* (Passeriformes, Fringillidae). *Journal of Eukaryotic Microbiology, 43,* 489-491.

Rysavy, B. (1954). Prispevek k poznani kokcidii nasich i dovezenych obratlovcu. *Ceskoslovenska Parasitologie, 1,* 131-174.

Scholtyseck, E. (1954). Untersuchungen uber die bei einheimischen vogelarten vorkommenden Coccidien der Gattung *Isospora*. *Archiv für Protistenkunde, 100,* 91-112.

Sick, H. (1997). *Ornitologia Brasileira.* Rio de Janeiro: Nova Fronteira, 862 pp.

Silva, E., Literák, I., & Koudela, B. (2006). Three new species of *Isospora* Schneider, 1881 (Apicomplexa: Eimeriidae) from the lesser seed-finch, *Oryzoborus angolensis* (Passeriformes: Emberizidae) from Brazil. *Memórias do Instituto Oswaldo Cruz, 101,* 573-576.

Soulsby, E. J. L. (1987). *Parasitología y enfermidades parasitárias*. México: Interamericana, 823 pp.

Stabler, R. M., & Kitzmiller, N. J. (1972). *Isospora petrochelidon* sp. n. (Protozoa: Eimeriidae) from the cliff swallow, *Petrochelidon pyrrhonota*. *Journal of Protozoology, 19,* 248-251.

Swayne, D. E., Getzy, D., Slemons, R. D., Bocetti, C., & Kramer, L. (1991). Coccidiosis as a cause of transmural lymphocytic enteritis and mortality in captive Nashville warblers (*Vermivora ruficapilla*). *Journal of Wildlife Diseases, 27*, 615-620.

Templar, A. C., McQuistion, T. E., & Capparella, A. P. (2004). A new coccidian parasite, *Isospora andesensis*, from the common bush tanager *Chlorospingus ophthalmicus* of South America. *Acta Protozoologica, 43,* 369-371.

Upton, S. J., Erst, J. V., Clubb, S. L., & Current, W. L. (1984). *Eimeria forresteri* n. sp. (Apicomplexa: Eimeriidae) from *Ramphastos toco* and a redescription of *Isospora graculai* from *Gracula religiosa*. *Systematic Parasitology, 6,* 237-240.

Upton, S. J., Current, W., & Clubb, S. (1985). Two new species of *Isospora* (Apicomplexa: Eimeriidae) from passeriform birds of South America. *Systematic Parasitology, 7,* 227-229.

Upton, S. J., Marchiondo, A. A., & Williams, R. N. (1988). New species of *Isospora* (Apicomplexa: Eimeriidae) from passeriform birds of Hawaii. *Systematic Parasitology, 12,* 81-85.

Upton, S. J., Langen, T. A., & Wright, T. F. (1995a). A new species of *Isospora* Schneider, 1881 (Apicomplexa: Eimeriidae) from the white-throated magpie jay *Calocitta formosa* (Passeriformes: Corvidae) from Costa Rica. *Systematic Parasitology, 31,* 195-199.

Upton, S. J., Stamper, M. A., & Whitaker, B. R. (1995b). *Isospora bellicosa* sp. n. (Apicomplexa) from a Peruvian red-breasted meadowlark, *Sturnella bellicosa* (Passeriformes: Icteridae). *Archiv für Protistenkunde, 145,* 132-134.

Upton, S. J., & Whitaker, B. (2000). New species of *Isospora* (Apicomplexa, Eimeriidae) from the troupial and white-edged oriole (*Icterus* spp.) (Aves, Passeriformes, Icteridae). *Acta Parasitologica, 45,* 67-70.

Upton S. J., Wilson, S. C., Norton, T. M., & Greiner, E. C. (2001). A new species of *Isospora* Schneider, 1881 (Apicomplexa: Eimeriidae) from the Bali (Rothschild's) mynah *Leucopsar rothschildi* (Passeriformes: Sturnidae), and comments concerning the genera *Atoxoplasma* Garnham, 1950 and Isospora. *Systematic Parasitology, 48,* 47-53.

Wobester, G. A., & Cawthorn, R. J. (1985). Exogenous and endogenous stages of *Isospora brachyrhynchi* sp. n. (Protozoa: Eimeriidae) from the American crow *Corvus brachyrhynchos* Brehm. *Canadian Journal of Zoology, 63,* 2639-2645.

In: Birds
Editors: Lucas Ruiz and Franco Iglesias

ISBN: 978-1-62948-104-3

Chapter 4

An Animal Geography of the Dominant Urban Avian Scavenger in Contrasting Case Studies

Michael O'Neal Campbell
Department of Geography, Thompson Rivers University, Kamloops, BC, Canada

Abstract

Animal geography and avian ecology represent two ways of looking at avian relations with the environment and human behavior. Animal geography is the geography of animals, including the examination of animals as actants; active agents creating relations with people as individuals and groups. Avian ecology is more concerned with statistical generalization, at inter- and intra- species levels. Urban areas provide instructive case studies for the application of such perspectives, due to the intense relation between animals and people in highly variable landscapes. In this chapter, case studies of the application of these paradigms and sub-disciplines are taken from published articles on urban birds in North America, Europe and Africa. In all the studies there were similar issues; dominant scavengers in relationship with smaller birds, and with people acting as non-feeders, and accidental and deliberate feeders. From the perspective of urban avian ecology, dominant avian scavengers obtained a statistically significant dominant relationship with smaller scavengers. From the perspective of animal geography, both dominant and dominated scavengers constructed individual strategies to enable food access. This individualism, in reaction to other birds, and towards variable, individualist human behaviors, could be defined as actancy, with evidence of innovation, experimentation and possibly strategic thinking. These studies illustrate the importance of flexibility in avian ecological studies.

Introduction

This chapter takes an animal geography perspective on the role of large avian scavengers - vultures, corvids and gulls - in ecological competition and nature society relations, in urban case studies from North America, Europe and Africa. Animal geography broadly considers the geography of animals, but recent strands of this sub-discipline have sought to examine animals as actants; active agents creating relations with people as individuals and groups. Actant behavior is defined as the adaptation to options for any reasons (Murdoch 1998). By contrast, avian ecology examines birds with statistical generalization, in relation to other animals and the environment.

I examine the interface of these two possibly complementary approaches, exploring the possibilities of both actant individualism and statistically significant behaviours in interspecies relations. Case studies are taken from four articles I published in 2007 – 2010 (see Campbell 2007, 2008, 2009, 2010). In the British case studies of Brighton and Glasgow, the dominant avian urban scavenger is the large gull; the lesser black backed gull, great black backed gull or the herring gull. In the Canadian case study of Peterborough, the dominant scavengers are ring billed gulls, but there is competition in some areas from the larger Canada geese, both being dominant over corvids, mallards ducks and doves. In the Ghanaian case study of Accra and Kumasi, the dominant avian urban scavengers are hooded vultures and pied crows. The case studies are compared in terms of evidence of actant individualism and behavioral generalization in dominant scavenger relations with other scavengers and people. The evidence shows substantial individualism that may be defined as actancy, especially in foraging and feeding space competition, including innovation, experimentation and possibly strategic thinking, in addition to statistically predictable behavior. This evidence transcends different species and human cultures in continental case studies, and underscores the value of a broad perspective in avian behavioral ecology. The rest of this chapter looks at the methodology of animal geography and examines how this perspective assists in the assessment of urban avian ecology.

Animal Geography

'Animal' geography, a framework for possibilities of active animal agency is 'a newly coined topic in human geography', promoting a more dynamic situation of animals, not merely as food resources or objects, but as active subjects (Philo and Wilbert 2000; Wolch and Emel 1998). Animal geography is a branch of geography that questions the dualism inherent in the contrasting paradigms of the statistically dependent ecological and more subjective social sciences. It aims to create a more critical and transparent engagement between these two opposites, which is constructed from relevant human geographies of animal-human relations and integrated space production (Campbell 2009). Animal geography's claim to utility is its emphasis on a more flexible framework than those of biogeography, ecology or environmental science. In this, some strands even allow the attribution of humanistic individualism to animals, including subjective and strategic behavior. Animals are examined, not as objects, but as subjects or actors exhibiting actancy (Murdoch 1998; Johnston et al. 2000). Such studies go beyond the simple impact of people on animals and resultant, generalizable animal adaptation. They seek to avoid biases arising from generalization, while simultaneously striving to avoid bias emerging from excessive individualism. This novel exploration is claimed to expose the limits of classical ecology and related sciences.

Traditional science has already developed important studies in animal – human relations (Lim and Sodhi 2004). An important testing ground for animal geography is the urban area. This is because animal – human relations are intensified, tested and modified in areas of high human population density, potential food sources dependent on human behavior, and artificial landcover that influence avian foraging change. Birds have been found to be an important, highly mobile urban animal group, relating to humans as deliberate and non - deliberate food providers (Belant, 1997; Burger, 1981; Clergeau et al. 1998; Cooper, 2002; Green and Baker 2003; Marzluff et al. 2001; Mitchell et al. 2004; Vuorisalo et al. 2003). An important question is: *To what extent would animal actancy emerge in such intensive contexts*?

Despite the important work of urban avian ecologists, the extremely variable socio-environmental relations in urban contexts, requires broader, more flexible study (Campbell 2007). Arguably, an 'important long term goal' for urban ecologists is to 'uncover the factors regulating the success or failure of species in inhabited areas, and use these factors to develop principles for the design of urban landscapes compatible with nature' (Turner 2003: 150). In particular, there must be a stronger focus on possibilities of animal and human agency and individualism than is emphasized in urban ecological and biogeographical studies. In classical ecology, there are few cases of intense human pressure where 'animals emerge as active agents fashioning the environment they inhabit or as experimental objects in the social relationships which bind them in different ways to people' (Johnston et al. 2000: 25).

The animal geography approach, when related to avian – human interactions in urban areas, recognises that: (1) birds occupy human spaces, with sufficient intensity to affect people in food provision, recreation landuse and land management; (2) people subjectively negotiate and renegotiate responses to avian participation in their life spaces, with continual reordering of such reactions; (3) by occupying intensively artificial human environments, birds constantly negotiate new life interactions, and are influenced by both the unique artificiality of the cityscape and the intense human pressure; and (4) therefore both humans and birds may be considered actants with a shared, evolving relationship. Both birds and people therefore exhibit possibly equal actant behavior; the adaptation of reactions for any reasons (Murdoch 1998; Campbell 2007, 2009, 2010). An animal geography with these emphases, complementing a positivist urban ecology, enables greater detail in avian studies.

The Methodology of Animal Geography and Avian Ecology

In the field studies described in this chapter, certain bird behaviors were recorded as representative of individual actancy, while other behaviors were statistically generalizable at the interspecies level. Individualism was recorded in

1 Waiting behaviors of birds prior to feeding, especially for human provided food.
2 The links between such behavior and individualist behavior in human feeders, especially when people adopted variable behavior.
3 Interspecies conflicts, especially evidence of strategic behavior in food access.

To measure individualist, actant behavior in birds, it was sometimes necessary to identify individual birds. Birds were identified by size, feather patterns, broken, dirty or missing feathers, and soiled feet, and these birds were observed for the different activities.

In addition, generalisable ecological behaviour was related to three main issues: (1) species presence according to land-cover variation and human presence/behaviour; (2) the human effects on bird presence; and (3) generalisable effects of people on bird behaviour.

To measure generalizable trends in birds, general patterns of behavior were based on food procurement activities, using flapping, pecking, chasing, dodging, hovering and moving around feeders. These translated into conflicts with winners and losers, the loser retreating from its previous feeding space and/or losing its food by dropping it or by violent removal from its bill. Also included were interspecies pursuits (on land, water or through the air) and physical attacks through pecking and/or wing flapping, and how these changed with variable human behaviors and food locations.

Both generalizable and individual bird behaviours were classified into three main periods. Firstly, there was the pre-feeding phase, which occurred before human presence, when birds were generally relaxed and perching, or quietly foraging in individual or group spaces. Secondly, there was the human presence phase, when birds clustered around non-feeding or feeding people, and interspecies and intraspecies conflicts emerged. Thirdly, there was the feeding and post feeding phase, when avian conflicts were linked to dropped food, and morsels dragged by fighting birds. There was also the post feeding phase, when most food was consumed, and birds were still in conflict, clustering around potential feeders, and contesting spaces at varying distances from the people.

Throughout these phases, bird numbers were recorded and statistical methods (correlation and regression) were used to analyze behavioral patterns. In such analyses, the dependent variable was the number of individual birds of each species. The independent variables were the number of human feeders, the distance of the feeders, the dominant landcover type, the number of birds of other species, the flight distances in metres, the number of inter- and intra-species interactions, and the distance from waterbodies.

Locations of the Case Studies

The case studies were taken from Accra and Kumasi in Ghana, Peterborough in Canada, and Brighton and Glasgow in the UK (see Campbell 2007, 2008, 2009, 2010). In continuity with these published articles, in this chapter animal geography works with ecological surveys, as it seeks to complement, rather than reject ecological methods and social surveys based on human behaviors that affect animals.

As already stated, relations between people and animals in intensive, densely populated, multi-species urban contexts are particularly relevant to such studies, because limited spaces increase contacts between actants, which may multiply the types of agency. For this reason, all my cases were located in urban areas near waterbodies (rivers or ocean), urban green spaces and multiple urban landuses (recreation, conservation, commercial, consumption, residential, industrial or vehicular. The multiple landuses ensured that people have different mindsets, intentions, perspectives and attitudes.

Three of the case studies are arguably more related, based in the temperate zone, in large urban centres with similar configurations (green spaces, built areas, car parks and waterbodies – rivers or sea). The fourth case, Accra and Kumasi does not involve seabirds or waterbirds, but shares all the other attributes.

Slight human cultural differences were apparent, as in the African case there were no recreational bird feeders in green spaces or streets, cancelling the classification in the other cases of human proffered food (HPF, food deliberately thrown by people to birds for feeding).

Glasgow (55.86 N, 4.26 W, population 598,830, National Records of Scotland 2012), the largest city in Scotland UK, has a reputation of a green city, due to the very large parks that dominate its area. It has been described as 'better endowed with green spaces than other European cities, these covering over 20 percent of its area' (Glasgow City Council 2004). It is bisected by the large Clyde river and smaller water bodies. A short distance from the Atlantic coast, it is strongly populated by seabirds. The commonest avian scavengers are the lesser black backed and black headed gulls, and mallard ducks (Figure 1). Smaller scavengers include the rock dove (Campbell 2007).

Brighton and Hove (50.84 N, 0.13 W, population 480,000 Office for National Statistics 2007), in the UK is a medium sized, coastal and very green city. Also a sanitary city, most of the food sources for avian scavengers are deliberate human feeds in green spaces, streets and beaches. A leading tourist city, it provides large recreational and outdoor eating spaces that encourage avian feeders.

English name	Scientific name	Length cm	Wingspan cm	Weight kg
Glasgow				
Lesser black-backed gull	*Larus fuscus*	52 - 64	135 - 150	0.6 - 1
Black-headed gull	*Larus ridibundus*	34 - 37	100 – 110	0.2 – 0.4
Mallard	*Anas platyrhynchos*	51 - 62	81 - 98	0.7 – 1.5
Rock dove	*Columba livia*	31-34	63 - 70	0.2 – 0.4
Brighton				
Herring gull	*Larus argentatus*	53 - 59	138 – 150	0.7 – 1.5
Lesser black-backed gull	*Larus fuscus*	52 - 64	135 - 150	0.6 - 1
Great black-backed gull	*Larus marinus*	64 - 78	150 - 165	1- 2
Black-headed gull	*Larus ridibundus*	34 - 37	100 - 110	0.2 – 0.4
Carrion Crow	*Corvus corone*	45 - 47	93 - 104	0.4 – 0.65
Rook	*Corvus frugilegus*	44 - 46	81 - 99	0.3 – 0.35
Jackdaw	*Corvus monedula*	33 - 34	67 - 74	0.2 – 0.25
Mallard	*Anas platyrhynchos*	51 - 62	81 - 98	0.75 – 1.5
Rock dove	*Columba livia*	31-34	63 - 70	0.2 – 0.4
Peterborough				
Ring billed gull	*Larus delawarensis*	43 - 54	105 - 117	0.3 – 0.7
Canada goose	*Branta canadensis*	56 - 110	122 - 183	0.4 - 0.55
Mallard duck	*Anas platyrhynchos*	51 - 62	81 -98	0.75 – 1.5
American Crow	*C.brachyrhynchos*	45	85-100	0.3 – 0.62
Starling	*Sturnus vulgaris*	22	37-42	0.06 – 0.09
Accra/Kumasi				
Rock dove	*Columba livia*	31 - 34	63	0.24 – 0. 4
Hooded vulture	Necrosyrtes monachus	62–72	155–165	0.5 - 2.6
Pied crow	Corvus albus	46–50	95 - 110	0.5 – 0.7

Figure 1. Birds in the combined studies.

The coastal location attracts several gull species, mainly herring, black headed, great and lesser black backed gulls, while the green hinterland attracts crows, rooks, jackaws, magpies, mallard ducks and many smaller species (Campbell 2010).

Peterborough (44.50 N, 78.50 W, population 74 698; Statistics Canada 2011), in Canada is located between Ottawa and Toronto, north of Lake Ontario. A fairly dispersed city, is has many large green spaces, and a large bisecting river. These conditions provide good conditions for urban avian scavengers. The main scavengers are ring billed gulls, Canada geese, mallard ducks and American crows (Figure 1). The distance to the coast is probably a limiting factor for seabird species. The good sanitary conditions and the absence of large rubbish dumps mean that avian scavengers are largely dependent on deliberate human feeds, small discards in green spaces and along river banks and street waste (Campbell 2008).

Accra (5.55 N, -0.2 W, population 2,291,352 Ghana Statistical Service 2012) and Kumasi (6.67 N, -1.62 W, 1,468,609, Ghana Statistical Service 2012), in Ghana are highly urbanized, with densely populated residential areas and numerous, waste producing urban markets, abattoirs and rubbish dumps. The sanitary conditions are generally poor. There are strong agricultural, industrial, transport and population links to the surrounding rural areas, and patchworks of farms and settlements overlap with the edges of the urban areas. Both cities are bisected by rivers and Accra is on the Atlantic coast. These conditions provide a good ecological setting for urban avian scavengers. *Larus* gulls are largely absent, the main avian scavengers being the hooded vulture and the pied crow. The hooded vulture is the smallest vulture in Ghana, but is the only vulture species in the southern half of the country (which includes Accra and Kumasi). The larger vulture species are found in the northern half of the country (Campbell 2009).

Relations between the Dominant Scavengers and Other Scavengers

This section considers the relations between the dominant avian scavenger species, other species and people in the case studies. Although the primary focus is on birds, human behavior is considered where it affects avian access to food, foraging behavior, inter- and intra- species conflicts and presence. Each case study is considered in turn, with the following topics: interspecies conflicts, from generalizable and actant related perspectives; and avian reactions to non-feeding, general feeding and selective feeding human presence.

The Case of Glasgow

Interspecies Conflict

The dominant avian scavenger was the lesser black backed gull. The other common scavengers were the rock dove and the mallard duck. Gulls dominated both ducks and doves, but ducks more so. Ninety five percent of the observed conflicts between gulls and ducks resulted in leaving behavior from ducks. The dominance of gulls over ducks was manifested by flapping, pecking, chasing and hovering. Ducks were robbed of food by retreating, and also had food morsels removed from their bills.

Gulls also dominated doves, with gulls winning ninety two percent of the observed conflicts. Dominance was manifested by wing flapping, as doves were generally too fast to be pecked. The superior size of the gulls was balanced by the agility of the doves. Ninety two percent of the doves were in constant movement, compared with 41 percent of the gulls. Gulls were outmaneuvered by doves in 46 percent of the conflicts. The balance between gull wing flapping, hovering and approaches, and dove mobility produced a statistically significant result in favour of the gulls, but weaker than the result for gulls over ducks.

Actant Behavior in Interspecies Conflict

Actant behavior was revealed by individual variations, sometimes specific to particular birds that varied from statistically valid results. The principal behaviors were occasional dove/duck groupings on gulls, doves sneaking under gulls to steal food, and aggressive individual doves and ducks prevailing over certain gulls. In two percent of the observed conflicts three and four doves including one identifiable by three toes and a few white wing feathers, were able to rob one to three gulls of food by running between the gulls' legs, while other members of the 'team' flapped around the gulls' heads. This technique was partly successful (about 30 percent of attempts), but gull 'victories' were statistically significant. No other observed doves employed this technique.

A commoner technique employed by these doves and others was for five or six doves to crowd onto one gull in possession of food. In some cases the doves were able to distract the gull, but in other cases the gull pulled the food and flew away, counterattacked (one young gull with dark feathers was particularly good at this), or several gulls grouped, fighting the doves and each other in a free for all. In such 'melees' individual skills were apparent, but gulls usually were more successful (not statistically significant).

Ducks were not as mobile as doves, and not small enough to hide under gulls. They did start group attacks on gulls on land and water. None of the land attacks were successful, as different gulls used their superior speed, aggression and mobility to prevail. Ducks employed three main strategies. Three or four ducks attacked a gull simultaneously, appearing to confuse it into dropping food; gulls reacted by increased flapping, calling and flight. Ducks also picked food and moved with it as a group. Gulls reacted individually, by flying over the ducks and picking the right moment to dive for the food, or by attacking individual ducks closest to the food, or by holding onto the food and flapping. Ducks reacted in some cases by sitting on the food as they would an egg, but gulls also attacked such ducks. Using individual identification techniques, these methods were found to be individual and not general.

On water, the ducks were more successful. Individuals would push their heads under water (while holding food in their bills), or cover the food on the surface with their wings and bodies. Gulls reacted diving at the ducks, and even sitting on the backs on the ducks. Gulls victories were statistically significant, but ducks prevailed in about a third of cases.

Bird-Human Relations

In terms of bird reactions to humans, there was both generalizable and individual behavior. Gull, duck and dove numbers increased with human presence even when there was no feeding. Bird numbers were positively connected with those of non-feeding and feeding humans. Human presence contributed to increased inter- and intra- species conflict, as the number of recorded avian conflicts were positively correlated with the number of people present.

Crucially, public opinion favoured ducks and doves over gulls. As 25 percent of people visited riverbanks, parks and concrete spaces to feed birds, a pro-duck or pro-dove feeding culture existed. This preferential feeding did not however change the gulls' dominance of feeding areas.

Rather, there appeared to be an increase in gull aggression to towards other birds. Gulls grouped on ducks or doves being fed and compelled them to give up the food, intercepted food in mid-air or even attacked people. In some cases, gull aggression 'forced' people to leave the feeding area.

Particular, individual strategies employed by gulls included:

1 Perching on people's heads.
2 Defecating on people (perhaps accidental).
3 Attacking people by flapping about their heads.
4 Attacking ducks and doves near the people.
5 Attacking ducks and doves distant from people, thereby preventing these birds from reaching feeding areas.

None of these behaviors were generalizable. Where individual birds were identifiable, certain individuals favoured one or two methods, while others merely perched or flew overhead and were not observed in such aggression. Both ducks and doves appeared to develop strategies to cope with increased gull aggression.

For ducks

1 When approaching feeding humans, ducks flew from the water surface rather than swam, to cope with the rapid approach of gulls.
2 Sitting between peoples' shoes or very close to people, to avoid gull attacks. Individual gull reactions ranged from leaving, hovering overhead, to attacking the human feeders.
3 Leaving the immediate vicinity and returning if or when the gull flock melted away.

Doves employed similar strategies and in addition

1 Where human feeders were on park benches, doves hid under the benches.
2 They came closer to people, even sitting on their shoes.
3 Certain doves adopted their circling, diving behavior of gulls, using their smaller size and greater speed to outwit gulls in some instances. Gull reactions created aerial conflicts, with individual skill being the deciding factor.

The Case of Brighton

Interspecies Conflict

The dominant scavenger was the herring gull. Lesser black backed and great black backed gulls were also present, but in much smaller numbers. The other common scavengers were black headed gulls, carrion crows, rooks, jackdaws, mallard ducks and rock doves. Herring gulls were the most aggressive scavengers, dominating ducks, corvids and doves.

Gull dominance over corvids was manifested by diving (when the corvid was on the ground), chasing (when the corvid was in the air), hopping and flapping (when both were on the ground) and pecking (when both were contesting for food). Herring gull dominance over gulls, ducks and doves was similar to that of the lesser black backed gull in Glasgow. The dominance of the herring gull over other species was statistically significant. However, during the field study at least, the numbers of ducks in Brighton were fewer and the number of doves much higher than in Glasgow. Herring gulls also appeared more numerous in Brighton than the lesser black backed gulls in Glasgow, possibly because of the coastal location.

Actant Behavior in Interspecies Conflict

Actant behavior was also recorded for all species. In conflicts between herring gulls and black headed gulls, herring gulls prevailed about 85 percent of the time, but innovative behavior of the smaller gull (greater agility, closer flights to people, faster reactions to feeders) enabled victories in the reminder of encounters. The herring gulls reacted by ducking from the fast moving smaller gull, tumbling in flight, diving to the water surface, or dropping the food and picking it up again, or simply bullying the smaller bird.

In conflicts between herring gulls and corvids, especially rooks and jackdaws in groups, the corvids showed individual coping strategies in the face of the dominant gulls. Individual corvids perched in trees, on roofs, and on the ground, while others flew in circles around the feeding gulls. From these vantage points, sneak attacks were common. Gulls reacted by attacking sitting corvids, even distant from food sources. Such attacks often resulted in long distance chases. No observations were made of corvids chasing gulls or other species. No conflicts were recorded between corvids and ducks. Corvids showed minor aggression towards doves, as a few individuals chased doves a few metres and retrieved their food. In such cases, doves showed only minor leaving behavior.

Bird- Human Relations

Similar to the Glasgow study, there was a positive correlation between bird and human numbers, even in non-feeding situations. Human presence was also positively correlated with avian conflicts, but only with a gull species present. In conflicts initiated by human feeding, herring gulls were predictably dominant over other species, even with deliberate, discriminatory feeding of other species. People preferred ducks and doves to gulls and corvids. Preferential feeding went to extremes, for example protecting ducks and doves in close proximity, and placing food inches away or right in their bills. Less direct methods were the cessation of feeding when gulls began to dominate, targeted tossing to preferred individuals, movement from one feeding spot to another, and arm waving and shouting in response to gull presence or aggression.

In reaction to these actions, gulls adopted individual strategies, from observational perching, hovering and diving for opportunities to bold attacks to drive other birds from feeders. Not all such methods were successful, as doves were faster and more mobile than gulls at the short range. Ducks were more easily dominated by gulls, but individual ducks were successful. Their actions included: eating food faster, anticipating the feeders throwing action and reaching to the feeders with open bills. Gulls also reacted by opening their bills at people at close range, and swooped at ducks sheltering near people.

Gull conflicts with corvids around people were slightly different, in that corvids were usually not preferentially fed. Gull strategies included observational perching or hovering, interception of food in mid-air, mid-air attacks on corvids, driving of corvids from feeding areas and attacks on people. These actions had limited success, but gulls were significantly more successful than corvids. Corvids reacted by dodging, perching (usually in trees with small branches were gulls would not perch) and flying around people. In such encounters, herring gulls were the most successful against carrion crows (mostly solitary), than the more social rook and least against the social, mobile jackdaws. Victories by herring gulls over the larger corvids were statistically significant, but those against jackdaws were not, mostly based on individual actions. The smaller black headed gull was more successful against the similarly mobile jackdaw, but slightly less so against the crow and rook. The results were not statistically significant, due to strong individual action.

The Case of Peterborough

Interspecies Conflict

The dominant scavengers were the ring billed gull, Canada goose, American crow and mallard duck. The lack of a discernible, predominant bird species was due to the smaller size of the ring billed gull (similar to the black headed gull), and the American crow (only a little larger than the jackdaw). The Canada goose is large and aggressive, but is not very mobile and contested for food only on water and riverbanks, not in mid-air or far from water. The behavior of the mallard duck and the rock dove were similar to those in the case studies of Glasgow and Brighton. In conflict situations, the strong point of the ring billed gull was aerial mobility, speed in the air and on the water surface, and aggression. The strong points of the crows were similar, but they never landed on water, and even avoided flying close to the surface. The rock dove was also very mobile, but avoided water. The strong point of the Canada goose was its unrivalled size, although sometimes this was insufficient against 'ganging up' strategies of gulls and the superior mobility of crows, gulls and doves.

The main competition was between geese and gulls, with a narrow edge of victory for the gulls (44.6%) compared to geese (34.4%), with many indecisive results (21%). The victories by gulls and geese over ducks were statistically significant. Gulls prevailed over ducks in statistically more cases than geese did, because of their superior speed and mobility, and their ability to fight in the air. While gulls chased ducks on land and water, and in the air for considerable distances, geese did so only on land and water for one or two metres. The numbers of gulls and geese were negatively correlated with the number of ducks near feeders, although ducks might congregate in the distance. Larger numbers of ducks did not affect the numbers of any other species.

Actant Behavior in Interspecies Conflict

The key actant behaviors concerned duck innovation against gulls and geese, and mobility issues between crows, gulls and doves. Ducks obtained only about 30 percent of food morsels, even when the sought morsel was closer to them than the marauding geese or gulls. Nevertheless, some individual ducks were moderately successful, being faster, more aggressive and more mobile, especially on the water surface.

Although increased gull or goose presence was negatively correlated with duck numbers, not all ducks left the scene under attack. Certain individuals, in some cases identifiable as repeat actors, stayed and were able to use strategies (usually sneak attacks on large food pieces being broken up) to penetrate flocks of gulls and geese. One such mallard, a male, was easily identifiable due to the feathers lost in such encounters. Other ducks also relied on food broken up in intra-species fights between mostly aerial gulls and geese in water.

When food broke up on the bank or in green spaces, crows and doves were also involved, and their superior mobility further robbed ducks of food. Crows and doves had few conflicts with ducks, but they did deprive ducks by diving on them from observational posts, snatching food and flying away, rather than attacking the bird itself. Once in the air, only gulls regularly challenged them, as ducks and geese rarely had aerial fights, despite a few incidents of ducks doing so.

In non-riverine areas, the main conflicts were between gulls, crows and doves. Both geese and ducks were fewer and appeared slower on land than water. As gulls were dominant, the battle was between particularly aggressive gulls, 'smart' crows and 'sneaky doves.' Individually aggressive gulls controlled areas by flying over the crowd of birds and diving at opportunity, including attacks on any birds feeding or near food, while other gulls indolently sat and watched. Certain crows were able to rob slowly moving gulls. Individualist behavior from doves meant that gulls were defeated on some occasions (less than 20 percent of the time), usually through tricks such as snatching of food from under their bodies, confusion in mass attacks by crows or doves, and fast escapes by these birds.

Bird-Human Relations

The numbers of all bird species were positively correlated to the number of people, even non-feeders. People preferred ducks, based on good opinions of these birds, while gulls and crows received only neutral or negative opinions. Gulls were seen as the smartest, as they arrived faster, and were more aggressive and 'pushy'. All species were seen as 'quite clever'. All respondents cited gulls as the smartest, arriving first and dominating much of the feeding. A human 'counter strategy' to the dominant strategies of gulls and geese (in reaction to preferential duck feeding) was to drop food on the ground closer to the feeder, to give ducks shelter and a better chance at food. This was initially successful, but eventually all the other birds moved closer and ducks were overpowered (geese on the ground, gulls on the ground and swooping close). Gull aggression included low, successful dives to remove food from people's hands, aerial defecation on human clothes, and flapping about people's heads. The aggressive behaviour of gulls towards the deliberately fed ducks contributed to people leaving the feeding area. People described gull behaviour as dangerous and cited this as a factor for their preference of ducks.

The Case of Accra and Kumasi

Interspecies Conflicts

The dominant avian scavengers were the hooded vulture and the pied crow. Between these two species, there were interspecies fights. Vultures outnumbered crows where there were large amounts of waste meat, such as butcheries and meat markets.

However, even in such areas they were less mobile than crows, perching in trees or roof tops, or circling slowly overhead. Crows walked, ran, hopped and invaded tables, doors and windows. Where there were interspecies conflicts over small or large food morsels, vultures were statistically dominant, because when they seized a morsel crows appeared unable to take it, and no vultures fled the crows.

By contrast, when crows seized a morsel and many vultures appeared the crows either continued to eat, flew away with bits of food, or left the scene with no food. Crows were faster than vultures in accessing food being thrown away as waste. Point count surveys also showed that they also foraged in places where they were few vultures, such as gutters, sidewalks, house windows, piles of old clothing, public toilets and under vehicles.

Actant Behavior in Interspecies Conflict

Actant behavior of the vultures and crows concerned principally their individualist abilities in procuring food from people. In the absence of recreational bird-feeding, this principally concerned the 'stealing' of food, or the invasion of human spaces to procure food. Crows were the main actants, as they were able to enter small windows, doors and gaps in roofs, and take valuable food from stakeholders.

Some individuals were adept at particular actions, entering eating places under the tables and taking food off tables when eaters were distracted. Others *were* shared benches with tolerant people, and even hopped onto the tables. These actions were highly individual, as most crows merely flew overhead or engaged in observational perching. Vultures were much less skilled, but some individuals were observed with similar behaviors in more open areas.

Bird - Human Relations

Bird numbers were positively correlated with human numbers where there was a possibility of food, and weakly positively correlated with human numbers where there was no immediate, apparent source of food.

Generally, people had good opinions of scavengers when they ate refuse and negative opinions when they ate valued food. In the meat production areas, male butchers and female meat buyers thought that vultures and crows helped reduce discards.

Nevertheless, to avoid thefts butchers stood close to meat cutting areas and in some cases had to drive vultures away. Crows were a bigger problem, because they were 'smarter' at sneaking past people, and they also ate vegetable matter (e.g. tomatoes and cassava intended for human consumption) in all markets. Crows were bolder and required hand clapping and shouting, and in some cases stone throwing to be controlled. In general markets with little meat, crows were often the only scavengers.

Individualist behavior in birds was frequently linked to spiritual powers. 'Clever' thefts, identifiable birds repeatedly visiting areas and watching people, birds perching on trees near scenes of recent accidents, deaths or graves, sleeping on roofs overnight and looking through windows led people to the belief that particular birds were spiritually possessed (with either the souls of the dead with malicious intent, or with demons).

Other individualist behavior was linked to the ability of birds, mostly crows, to get food. People rarely fed birds deliberately, but they threw food at birds in markets, food shops and open air resturants, and noticed particular crows were 'bad' or 'smart' at accessing the food. In some cases, crows harassed food throwers, and even sat on them.

Conclusion

How do these issues contribute to our understanding of shared animal – human actancy? How do these studies contribute to the work of the urban avian ecologist? And how does this study illustrate the field situation of the animal geographer, as a person with a perspective requiring more critical perceptions of animal and human behavior, and the evaluation of both of these dynamics? The answers to these questions concern the extent to which this methodology has shown a way beyond the vocabulary of generalisable science, and looked at behavioral adaptation and even opportunism, opening a vision of more subjective bird behavior, co-dependent with human actions, as complementary rather than hostile to classical ecology. The studies showed remarkable similarities, yet there were still strongly individual behaviors from the studied species. The results hover between the need for scientific generalization, and the value of individual detail. Similar to the social sciences, there must be avoidance of the polarization that emerges between the generalizations of structural analyses and the focus of actor oriented studies, with acknowledgement of the strengths and weaknesses of both methods.

References

Aberra, E. and King, R. 2005. *Additional knowledge of livelihoods in the Kumasi Peri Urban Interface (KPUI)*. Ashanti Region Ghana University College, London.

Adams, L. W. 1994. *Urban Wildlife Habitat: A Landscape Perspective*. University of Minnesota Press, Minneapolis.

Adeola, M. O. 1992. Importance of wild animals and their parts in the culture religious festivals and traditional medicine in Nigeria. *Environmental Conservation* 19: 25–134.

Ankney, C. D., Dennis, D. G. and Bailey, R. C. 1987. Increasing mallards, decreasing American black ducks: coincidence or cause and effect? *Journal of Wildlife Management* 51: 523–29.

Asdal, K. 2003. The problematic nature of nature: the post constructivist challenge to environmental history. *History and Theory* 42: 60–74

Austin, J. E. 1988. Winter ecology of Canada geese in north central Missouri. *Unpublished PhD thesis*. University Missouri, Columbia MO

Bartelt, G. A. 1987. Effects of disturbance and hunting on the behaviour of Canada goose family groups in east central Wisconsin. *Journal of Wildlife Management* 51: 517–22.

Belant, J. L. 1997. Gulls in urban environments: landscape-level management to reduce conflict. *Landscape and Urban Planning*, 38: 245–258.

Bennett, C. F. 1960. Cultural animal geography: an inviting field of research. *Professional Geographer*, 12: 12–14.

Blair, R. B. 1996. Land use and avian species diversity along an urban gradient. *Ecological Applications*, 6: 506–519.

Borrow, N. and Demey, R. 2001 *The Birds of Western Africa*. Christopher Helm, London

Brooks, T. and Thompson, H. S. 2001. Current bird conservation issues in Africa. *The Auk* 118: 575–582.

Burger, J. 1981. Feeding competition between Laughing Gulls and Herring Gulls at a sanitary landfill. *Condor* 83: 328–335.

Burger, J. and Gochfield, M. 1998. Effects of eco-tourists on bird behaviour at Loxahatchee National Wildlife Refuge, Florida. *Environmental Conservation*, 25: 13–21.

Campbell, M. 2006. Urban parks shared spaces? The utility of alert distances as indicators of avian tolerance of human in the Stirling Scotland. *Area* 38 301–11

Campbell, M. 2007. An animal geography of avian ecology in Glasgow. *Applied Geography* 27: 78–88.

Campbell, M. 2008. An animal geography of avian feeding habits in Peterborough. *Area* 40 (4): 472 – 480.

Campbell, M. 2009. Factors for the presence of avian scavengers in Accra and Kumasi. *Area* 41(9): 341 – 349.

Campbell, M. An animal geography of avian foraging competition on the Sussex coast of England. *Journal of Coastal Research* 26(1): 44 – 52.

Canagarajah, S, Newman, C. and Bhattamishra, R. 2001. Nonfarm income gender and inequality: evidence from rural Ghana and Uganda. *Food Policy* 26 405–20.

Chace, J. F., Walsh, J. J., Cruz, A., Prather, J. W., and Swanson, H. M. (2003). Spatial and temporal activity patterns of the brood parasitic brown headed cowbird at an urban/ wildland interface. *Landscape and Urban Planning*, 64, 179–190.

Clergeau, P., Savard, J. P. L., Mennechez, G., Sauvage, A., and Falardeau, G., 1998. Bird abundance and diversity along an urban-rural gradient: a comparative study between two cities on different continents. *Condor*, 100, 413–425.

Cooper, D. S. 2002. Geographic associations of breeding bird distribution in an urban open space. *Biological Conservation*, 104, 205–210.

Emel, J., Wilbert, C. and Wolch, J., 2002. Society and animals. *Journal of Human-Animal Studies,* 10(4), 1.

Galvan, I. 2003. Intra-specific kleptoparasitism in Lesser Black Backed Gulls wintering inland in Spain. *Waterbirds*, 26(3), 325–330.

Gbogbo, F. and Awotwe-Pratt, V. P. 2008. Waste management and Hooded Vultures on the Legon Campus of the University of Ghana in Accra, Ghana. *West Africa Vulture News* 58 16–22

Glasgow City Council. 2004. Wildlife news: Conservation action. Glasgow: GCC.

Gough, K. V. and Yankson, P. W. K. 2000. Land markets in African cities: the case of peri-urban Accra. *Ghana Urban Studies* 37 2485–500.

Green, D. M. and Baker, M. G. 2003. Urbanisation impacts on habitat and bird communities in a Sonoran desert ecosystem. *Landscape and Urban Planning*, 63, 225–239.

Hebert, P. N. and McNeil, R. 1999. Nocturnal activity of ringbilled gulls at and away from the colony. *Waterbirds* 22 445–51

Hume, R. 2004. *Complete birds of Britain and Europe*. London: Dorling Kindersley Ltd.

Johnston, R. J., Gregory, D., Pratt, G., and Watts, M. 2000. *Human geography*, Blackwell, Malden.

Jokimaki, J. 1999. Occurrence of breeding bird species in urban parks: Effects of park structure and broad-scale variables. *Urban Ecosystems*, 3: 21–34.

Lim, H. C. and Sodhi, N. S. 2004. Responses of avian guilds to urbanisation in a tropical city. *Landscape and Urban Planning*, 66(4), 199–215.

Lima, S. L. and Zollner, P. A., 1996. Towards a behavioural ecology of ecological landscapes. *Trends in Ecology and Evolution*, 11, 131–135.

Marzluff, J. M., Bowman, R. and Donnelly, R. eds 2001 *Avian ecology and conservation in an urbanising world.* Kluwer Academic, Boston MA

Maver, L. (1998). Glasgow's public parks and the community 1850–1914: A case study in Scottish civic interventionism. *Urban History*, 25(3), 323–347.

Mavor, R. A., Parsons, M., Heubeck, M., Pickerell, G., and Schmidt, S. 2003. Seabird numbers and breeding success in Britain and Ireland. *2002 Joint Nature Conservation Committee*, Peterborough

Mitchell, I. P., Newton, S. F., Radcliffe, N., and Dunn, T. E. 2004. *Seabird Populations in Britain and Ireland.* T. and A.D. Poyser, London.

Moore, J. E. and Switzer, P. V. 1998. Pre-roosting aggregations in the American Crow Corvus brachyrhynchos. *Canadian Journal of Zoology* 76: 508 –512.

Mundy, P., Butchart, D., Ledger, J., and Piper, S. 1992. *The vultures of Africa*, Acorn Press, Russel Friedman Books, Johannesburg.

Murdoch, J. 1998. The spaces of actor network theory. *Geoforum*, 29(4): 357–374.

Ntiamoa-Baidu, Y., Owusu, E. H., Asamoah, S., and Owusu-Boateng, K. 2000. Distribution and abundance of forest birds in Ghana. *Ostrich* 71: 262–268.

Obosu-Mensah, K. 1999. *Food production in urban areas: a case study of urban agriculture in Accra*, Ghana Ashgate, Aldershot.

Park, C. R. and Lee, W. S. 2000. Relationship between species composition and area in breeding birds of urban woods in Seoul Korea. *Landscape and Urban Planning* 51: 29–36.

Philo, C. and Wilbert, C., 2000. *Animal Spaces, Beastly Places: New Geographies of Human-Animal Relations*. Routledge, London.

Philo, C., 1998. Animal, geography and the city: notes on inclusions and exclusions. In: Wolch, J. and Emel, J. (eds.), *Animal Geographies: Place, Politics and Identity in the Nature-Culture Borderlands*, 51 – 71. Verso, London.

Raven, S. J. and Coulson, J. C., 1997. The distribution and abundance of Larus gulls nesting on buildings in Britain and Ireland. *Bird Study*, 44: 13–34.

Rock, P. 2002. Roof Nesting Gulls in Bath. Follow Up Survey Conducted in April 2002 and Summarising Surveys from 1995. *Report to the Department of Environmental Services*. Bath: Northeast Somerset Council.

Rosenzweig, K. V., 2003. *Win-Win Ecology: How Earth's Species Can Survive in the Midst of Human Enterprise.* New York: Oxford University Press.

Savard, J. P. L., Clergeau, P. and Mennechez, G., 2000. Biodiversity concepts and urban ecosystems. *Landscape and Urban Planning*, 48: 131–142.

Schochat, E., Warren, P. S. and Faeth, S. H. 2006. Future directions in urban ecology. *Trends in Ecology and Evolution* 21 661–2

Traut, A. H. and Hostetler, M. E. 2004. Urban lakes and waterbirds: Effects of shoreline development on avian distribution. *Landscape and Urban Planning*, 69, 69–85.

Turner, W. R. 2003. City biological monitoring as a tool for ecology and conservation in urban landscapes: The case of the Tucson bird count. *Landscape and Urban Planning* 65: 149–166.

Vuorisalo, T., Andersson, H., Hugg, T., Lahtinen, R., Laaksonen, H., and Lehikoinen, E. 2003. Urban development from an avian perspective: causes of hooded crow Corvus corone cornix urbanisation in two Finnish cities. *Landscape and Urban Planning* 62 69–87

Whatmore, S. 1999. Hybrid geographies: rethinking the 'human' in human geography. In: Massey, D., Allen, J. L. and Sarre, P. (eds) *Human geography Today*, 22 – 40. Polity Press, Cambridge.

Whatmore, S. 2002. *Hybrid geographies: natures, cultures, spaces*. Routledge, London

White, J. G., Antos, M. J., Fitzsimons, J. A., and Palmer, G. C. 2005. Non-uniform bird assemblages in urban environments: the influence of street vegetation. *Landscape and Urban Planning*, 71, 123–135.

Wolch, J. and Emel, J. 1998. *Animal Geographies: Place, Politics and Identity in the Nature-Culture Borderlands*. Verso, London.

In: Birds
Editors: Lucas Ruiz and Franco Iglesias

ISBN: 978-1-62948-104-3

Chapter 5

Recent Infectious Diseases or their Responsible Agents Recorded from Japanese Wild Birds

***Takuro Hirayama*[1], *Kii Ushiyama*[1], *Yuichi Osa*[2] and *Mitsuhiko Asakawa*[*1, 3]**

[1]Department of Pathobiology, School of Veterinary Medicine, Rakuno Gakuen University, Ebetsu, Hokkaido, Japan
[2]Eastern Hokkaido Wildlife Station, Institute of Environmental Science, Hokkaido Research Organization, Kushiro, Hokkaido, Japan
[3]Wild Animal Medical Center, Graduate School of Veterinary Medicine, Rakuno Gakuen University, Ebetsu, Hokkaido, Japan

Abstract

Most infectious micro agents, e. g., parasitic and/or commensal viruses, bacteria, fungus and protozoans, are component parts of the ecosystems in which their wild avian hosts occur, they have co-evolved, and the infectious agents do not necessarily cause disease. Sometimes, however, ten of thousands of birds become sick and in extreme cases die, in various parts of the world because of an outbreak of an epidemic disease. By the way, considerable environmental change has occurred in Japan over the last 150 years as well. Furthermore, Japan is an important transit country for a wide range of migratory species. Japan is situated on the East Asian Flyway a migratory route, connecting northeast Asia with Southeast Asia, with important branches passing through the Nansei Shoto, Kyushu, Honshu and Hokkaido into northeast Russia and also via Kyushu and the Korean Peninsula into eastern China. The outbreaks of infectious disease have not so far significantly impacted populations of birds in Japan. Because an understanding of the infectious avian disease organisms that are present in Japan, their ecology, transmission, and symptoms is an essential conservation tool, we have overviewed this field by the end of the 20 century. However, according to the decade passed by, several issues have to be

[*] Correspondence author: askam@rakuno.ac.jp.

added to the review. Hence, this chapter provides a recent overview of the infectious diseases and/or their responsible micro agents recorded in Japan, and briefly potential strategies for epidemic risk reduction are given. For example, in viral diseases or agents, avian pox, rotaviral enteritis or arthritis, Newcastle disease, avian paramyxovirus infection, highly pathogenic avian influenza and West Nile fever were mentioned, respectively. Especially, since the WNV has been reported from the Far East, the virus will invade to Japan in nearly future. According to the computer simulation of the risk assessment of extinction due to an outbreak of West Nile disease among free ranging Red-crowned Crane (*Grus japonensis*) in Hokkaido Island, Japan, with usage of the software "Vortex", the population will die out within 40 years. Hence, the VecTest has been performed on oral and tissue swabs from all wild birds carried to the Wild Animal Medical Center in the Teaching Animal Hospital of the School of Veterinary Medicine, Rakuno Gakuen University. In bacterial diseases epidemiology, systematic surveillance and/or individual case reports on *Esherichia coli, Actinomyces* spp., *Plesiomonas shigelloides, Staphylococcus epidermidis Salmonella enterica* serotype Typhimurium DT40, *Vibrio cholera, V. parahaemolyticus, Campylobacter jejuni, Chlamydophila psittaci, Mycobacterium genavense* etc were given. Above all, mass mortality of sparrows (*Passer montanus*) due to an epidemic caused by *S. enterica* has been just started from 2005 to now. And, the mortality has seemed to be often associated with a chemical deicer poisoning and/or staphylococcal infection, respectively. Although diseases due to biotoxins are defined as a non-infectious disease, some of the diseases have an intimate relationship with presence of certain responsible bacteria, e. g., cyanobacteria, such as the genera *Microcystis, Anabaena, Aphanizomenon, Nodularia*, and *Oscillatoria*. And, some suspicious cases of such problems were recorded even in Japan. Several fungal genera, in particular *Candida, Cryptococcus, Aspergillus, Microsporum, Trichophyton, Fusarium, Ochroconis* and *Absidia* have been reported as causing diseases as well, and some suspicious cases in free-ranging birds without a positive identification of each responsible fungal species have been given. Furthermore, there are several genera of avian flagellates and amoeba (Phylum Sarcomastigophora), such as the genera *Trypanosoma, Hexamita, Histomonas, Parahistomonas, Monocercomonas, Trichomonas, Tetratrichomonas, Chilomastix, Entamoeba, Endolimax* etc. In particular, several cases due to *Trichomonas, Tetratrichomonas* and/or *Hexamita* infection have been found in wild birds, but there have been no reported mass mortality due to the protozoan agents in Japan. And, *Trypanosoma* spp. have, however, been found in wild and/or captive passerines and owls, but they seem to have been non-pathogenic, at least now. On the other hand, some of avian coccidian and haematozoan genera (Phylum Apicomplexa), such as *Eimeria, Isospora, Tyzzeria, Wenyonella, Caryospora, Cryptosporidium, Sarcocystis, Toxoplasma, Atoxoplasma, Hepatozoon, Haemoproteus, Leucocytozoon, Plasmodium* and *Babesia*, may be responsible for severe disease. First of all, several cases including observatrion of *Isospora* shaped oocysts of *Atoxoplasma* sp. from fecal examinations and/or histopathological observation from postmortem examinations from not only captive Bali Myna (*Leucopsar rothschildi*), but also free-ranging passerines, e.g., *Passer montanus, Chloris sinica* and *Spodiopsar cineraceus.* There have been many case reports on intestinal coccidiosis including *Cryptosporidium meleagridis* infection from common avian species, it is prominent that *Eimeria ueki* was found from endangered subspecies, *Lagopus muta japonica*, on high mountain area. In 2002, fatal cases of the Red-crowned cranes due to disseminated visceral coccidiosis, caused by *Eimeria reichenowi* and *E. gruis*, were found. Control of the visceral coccidian species is more difficult than the control of intestinal coccidian species. Adding to this, we could observe renal coccidiosis concomitant infection with *Corynebacterium* sp. and *Renicola* sp. from mass mortality of *Puffinus tenuirostris. Sarcocystis*, most commonly found among waterfowl in the United States, is nonfatal, but severe infections can cause loss of muscle tissue and result in lameness, weakness etc. Recently, there have been several records of the protozoans among wild birds in Japan as well. Within the areas of both public health and captive animal health, toxoplasmosis among poultry is also very important. Serological evidence of the protozoan infection has been found so far among almost one third individuals of free-ranging anatids in Obihiro, Hokkaido Island. Although the agents for avian malaria including the genera *Leucocytozoon*, *Haemoproteus* and *Plasmodium* have been commonly found among various avian species including endangered ones, e.g., *Ketupa blakistoni, Haliaeetus pelagicus, Lagopus muta japonica* etc and exotic one, e.g., *Leiothrix lutea*, in Japan. Adding to the protozoans, at least in Japan, hosts of the genus *Babesia* have been restricted to mammals, but the piroplasma was obtained from *Spilornis cheela* in Yaeyama Archipelago, southern part of Japan. Certain free living protozoans such as diatoms (Phylum Bacillariophyta: genera *Nitzschia* and *Pseudonitzschia*) and dinoflagellates (Phylum Dinoflagellata: genera *Alexandrium* and *Gymnodinium*) produce neurotoxins such as domoic acid, saxitoxin, and brevetoxin, that can affect sea birds. Most reports of mortality in birds have been die-offs occurring in conjunction with protozoan blooms in the United States, but no such mortalities have been reported yet in Japan. Nevertheless, many protozoan blooms occur around Japan making it essential that we pay attention to such events because of their potential for causing mortality. Including the diseases, the disease outbreak or mass mortality prevention will need to be based not only on an understanding of avian ecology but also of agents (pathogens) epidemiology. And, in order for medical researchers to be able to prepare for future epidemic outbreaks amongst birds, they must perform epidemiological surveys of the probable agents mentioned above in wild bird populations both in Japan and its surrounding areas; international and inter-agency cooperation is, therefore, essential in disease prevention. Needless to say, very careful attention should be paid to the possibility of infection by zoonotic agents during monitoring surveys.

Introduction

Most infectious micro agents, namely, parasitic and/or commensal viruses, bacteria and protozoans, are component parts of the ecosystems in which their wild avian hosts occur, they have co-evolved, and the infectious agents do not necessarily cause disease. Sometimes, however, ten of thousands of birds become sick and in extreme cases die, in various parts of the world because of an outbreak of an epidemic disease (Friend and Franson, 1999).

By the way, considerable environmental change has occurred in Japan over the last 150 years as well (Asakawa et al., 2013). Furthermore, Japan is an important transit country for a wide range of migratory species. Japan is situated on the East Asian Flyway a migratory route, connecting northeast Asia with Southeast Asia, with important branches passing through the Nansei Shoto, Kyushu, Honshu and Hokkaido into northeast Russia and also via Kyushu and the Korean Peninsula into eastern China (Committee for Check-List of Japanese Birds of Ornithological Society of Japan, 2012). The outbreaks of infectious disease have not so far significantly impacted populations of birds in Japan.

Because an understanding of the infectious avian disease organisms that are present in Japan, their ecology, transmission, and symptoms is an essential conservation tool (Friend and Franson, 1999), we have overviewed this field by the end of the 20 century (Asakawa et al., 2002). However, according to the decade passed by, several issues have to be added to the review paper. Hence, this chapter provides a recent overview of the infectious diseases and/or their responsible micro agents (e. g., virus, bacteria, fungi and protozoa) recorded in Japan, and briefly potential strategies for epidemic risk reduction are given.

Viral Diseases

There have been recent cases of the avian pox viral disease in several species including *Haliaeetus albicilla, Milvus migrans, Columba livia, Emberiza schoeniclus, Corvus corone, C. macrorhynchos* and *Passer montanus* (Kubo et al., 2005; Yamaguchi et al., 2008; Saito et al., 2009; Watanabe et al., 2010; Fukui et al., 2013). Among them, an infection of the cases of *Corvus* spp. seemed to be related with garbage from town people (Fukui et al., 2013). Because there has been a fatal case of *Anser albifrons* infected by Marek's disease virus in Hokkaido (Asakawa et al., 2013), the authors have performed postmortem examinations on 12 individuals of the goose species between 2002 and 2010, but there are no further such cases (Nakamura et al., 2004; Yoshino and Asakawa, 2012). Anyway, a brief review on recent viral diseases of anatids in Japan has been published (Asakawa et al., 2013), so the data will be excluded from here.

As RNA virus recorded recently, avian rotavirus was detected from *Phalacrocorax carbo* and *Columba livia* (Suzuoka et al., 2002; Sugiyama et al., 2006), but mass mortality due to the rotaviral enteritis or arthritis have not occurred. Newcastle disease virus is capable of infecting a wide range of avian species. In addition to domestic poultry, amongst which it is particularly virulent, more than 230 species of birds from more than half of the avian orders have been found to be susceptible to natural or experimental infections with the virus. Hence, a systematic epidemiological survey has been done between 2001 and 2007 (Jahangir et al., 2009; Mase et al., 2009). Adding to this, the survey for the avian paramyxovirus has been performed as well (Fujimoto et al., 2010).

Recently, wild birds have been the focus of concern as a source for highly pathogenic avian influenza (abbreviated to HPAI) virus infections in Japan, because an outbreak of the HPAI has been occurred since 2004 in the country. So, systematic surveys have been done, and the HPAI virus has been isolated from total 60 individuals belonging to 15 species of wild birds which were not only migratory, but also were resident (Jahangir et al., 2009; Fujimoto et al., 2010; Okamatsu et al., 2010; Yamamoto et al., 2011; Bui et al., 2012; Uchida et al., 2012; Neagari, 2013).

In the mosquito-borne flaviviruses, West Nile virus (abbreviated to WNV) has been regarded as one of the most emerging zoonotic agents, and antibodies to the viruses were positive in mainly wild birds in Japan (Saito et al. 2009, 2011, Uchida et al., 2012). Although tests used with mosquitoes and birds for the viruses themselves and/or WNV RNA provided no evidence until now, the antibody was sero-positive in several individuals (Saito et al. 2009, 2011). And, since the WNV has been reported from the Far East (Saito et al. 2009, 2011), the virus will invade to Japan in nearly future. According to the computer simulation of the risk assessment of extinction due to an outbreak of West Nile disease among free ranging Red-crowned Crane (*Grus japonensis*) in Hokkaido Island, most northern part of Japan, with usage of the software "Vortex", the population will die out within 40 years (Onuma et al., 2010). Hence, the VecTest (Medical Analysis Systems, Inc., USA; antigen-capture assay for West Nile virus) has been performed on oral and tissue swabs from all wild birds carried to the Wild Animal Medical Center (Asakawa, 2010; abbreviated to WAMC) in the Teaching Animal Hospital of the School of Veterinary Medicine, Rakuno Gakuen University (abbreviated to RGU), Japan, since 2006 (Asakawa, 2006, 2008; Yoshino and Asakawa, 2012).

Bacterial Diseases

Epidemiologic surveillance of *Esherichia coli* has been performed on sysatematic collection of avian species including *Treron sieboldii, Egretta garzetta, Puffinus tenuirostris.* Grus spp., *Spodiopsar cineraceus, Emberiza cioides, Cyanopica cyanus, Corvus* spp. etc, because of several outbreaks of enterohemorrhagic *E. coli* infection among Japanese human society (Fukuyama et al., 2003; Asaoka et al., 2004; Kokayashi et al., 2009; Kitadai et al., 2013). However, no clear evidence of wild birds to transport the agents was found out. But, Asaoka et al. (2004) could find *Actinomyces* spp. and *Plesiomonas shigelloides* from *Corvus macrorhynchos*, and Yamaguchi et al. (2003) did *Staphylococcus epidermidis* from *Lagopus muta japonica,* respectively.

Mass mortality of sparrows (*Passer montanus*) has been due to an epidemic caused by Salmonella enterica serotype Typhimurium DT40 (Kobayashi et al., 2007; Fukui et al., 2009; Fujii and Osa, 2013) since 2005. And, the mortality has seemed to be often associated with a chemical deicer poisoning (Tanaka et al., 2008) and/or staphylococcal infection (Asakawa, 2006), respectively. On the other bacterial agents responsible for the infectious diseases and/or mass mortality due to the agents, fatal necrotic enteritis associated with Clostridium perfringens in crows (Asaoka et al., 2004), *Vibrio cholera, V. parahaemolyticus, Campylobacter jejuni, C. coli, C.* sp. and *Helicobacter* sp. obtained from several water birds, black kites and feral pigeons or doves (Watanabe et al., 2003; Kawamori et al., 2004; Hayakawa et al., 2011), respectively.

Because of human pneumonia, both genera *Chlamydophila* and *Mycobacterium* are regarded as public concern. Hence, Ohnishi et al. (2006) reported *C. psittaci* from feces of *Columba livia, Hypsipetes amaurotis* and *Strix uralensis* with the PCR method. Furthermore, mass mortality of *Geopelia cuneata* due to *Mycobacterium* genavense infection (Fukuda et al., 2008).

Although diseases due to biotoxins (poisoning) is defined as a non-infectious disease sense strico, some of the diseases have an intimate relationship with presence of certain responsible bacteria, e. g., cyanobacteria, such as the genera *Microcystis, Anabaena, Aphanizomenon, Nodularia* and *Oscillatoria*, that produce hepato- and/or neurotoxins affected a wide range of avian species (Asakawa et al., 2002). And, some suspicious cases of such problems were recorded in Japan as well (Fukui, 2013; Neagari, 2013).

Fungal Diseases

Several fungal genera, in particular *Candida, Cryptococcus, Aspergillus, Microsporum, Trichophyton, Fusarium, Ochroconis* and *Absidia* have been reported as causing diseases including large-scale mortality events and/or mycotoxicosis in a range of avian species from various bird species, and many cases including fatal ones in zoological gardens and aquariums in Japan (Asakawa and Kitayama, 2003). And, some suspicious cases in free-ranging birds without a positive identification of each responsible fungal species have been given (Ishimoto et al., 2007; Asakawa, unpubl.).

Parasitic Protozoan Diseases

There are several genera of avian flagellates and amoeba (Phylum Sarcomastigophora), such as the genera *Trypanosoma, Hexamita, Histomonas, Parahistomonas, Monocercomonas, Trichomonas, Tetratrichomonas, Chilomastix, Entamoeba, Endolimax* etc (Asakawa et al., 2002). Among them including a report of postmortem examination without positive protozoan identification, several cases due to *Trichomonas, Tetratrichomonas* and/or *Hexamita* infection have been found in wild and captive birds (Ishimoto et al., 2007; Nishimura et al., 2009), but there have been no reported mass mortality due to the protozoan agents in Japan. And, *Trypanosoma* spp. have, however, been found in wild and/or captive passerines (Passeriformes) and owls (Strigiformes) (Nagata and Sodhi, 2003; WAMC, unpubl.), but they seem to have been non-pathogenic.

On the other hand, some of avian coccidian and haematozoan genera (Phylum Apicomplexa), such as *Eimeria, Isospora, Tyzzeria, Wenyonella, Caryospora, Cryptosporidium, Sarcocystis, Toxoplasma, Atoxoplasma, Hepatozoon, Haemoproteus, Leucocytozoon, Plasmodium* and *Babesia*, may be responsible for severe diseases (Friend and Franson, 1999; Asakawa et al., 2002). First of all, although no report of the genus *Atoxoplasma* was found before Asakawa et al. (2002), several cases including observatrion of *Isospora* shaped oocysts of *Atoxoplasma* sp. from fecal examinations and/or histopathological observation from postmortem examinations from not only captive for re-introduction in Yokohama Zoo, e.g., Bali Myna (*Leucopsar rothschildi*; Sato et al., 2004), but also free-ranging passerines, e.g., *Passer montanus, Chloris sinica* and *Spodiopsar cineraceus* (Ueda et al., 2003; Kubo et al., 2005; Fukui et al., 2006). There have been many case reports on intestinal coccidiosis including *Cryptosporidium meleagridis* infection from common avian species (Kubo et al., 2005; Fukui et al., 2006; Kuroki et al., 2006; Nishimori et al., 2009), it is prominent that *Eimeria ueki* was found from endangered subspecies, *Lagopus muta japonica*, on high mountain area of Honshu Island, Japan (Shiibashi et al., 2005; Sato, 2011).

In 2002, fatal cases of the Red-crowned cranes, *Grus japonensis*, due to disseminated visceral coccidiosis, caused by *Eimeria reichenowi* and *E. gruis*, were found (Watanabe et al., 2003). This was memorable as well, because it was first free-ranging case of the diseases in the cranes designated as a special natural treasure by the Japanese Government. So, an emergent epidemiological surveillance was performed, and the oocysts were detected from many crane individuals (Sarashina et al., 2006; Ueda et al., 2008; Honma et al., 2011; Matsubayashi et al., 2005). Control of the visceral coccidian species is more difficult than the control of intestinal coccidian species, since the levels of anticoccidial drugs given orally may be lowered prior to reaching the site of infection such as the liver, muscles, spleen, and lung, both by low absorption into the tissue and detoxification by the host (Asakawa et al., 2002). Adding to this, Asakawa et al. (2008) could observe renal coccidiosis concomitant infection with *Corynebacterium* sp. (Bacteria) and *Renicola* sp. (Trematoda) from 1 individual of mass mortality of *Puffinus tenuirostris.*

Sarcocystis, most commonly found among waterfowl in the United States, is nonfatal, but severe infections can cause loss of muscle tissue and result in lameness, weakness, and even, in rare cases, paralysis (Friend and Franson, 1999). Because the debilitating effects of severe infections among birds could increase their susceptibility to predation and/or to other causes of mortality, it is important that we should know of the natural distribution of localities with identification of its final hosts for the conservation biology of the avian species. And, recently, there have been several records of the protozoans among wild birds (e. g., anatids, larids and passerins) in Japan (Kubo et al., 2006; Ejiri et al., 2008). Within the areas of both public health and captive animal health, toxoplasmosis among poultry is also very important. Serological evidence (ca. 30 %) of the protozoan infection has been found so far among free-ranging anatids in Obihiro, Hokkaido Island (Murao et al., 2008).

Although the agents for avian malaria including the genera *Leucocytozoon*, *Haemoproteus* and *Plasmodium* have been commonly found among various avian specis including endangered ones, e.g., *Ketupa blakistoni, Haliaeetus pelagicus, Lagopus muta japonica* etc and exotic one, e. g., *Leiothrix lutea*, in Japan (Nagata and Sodhi, 2003; Nakayama and Isobe, 2004; Hisada et al., 2004; Hagiwara et al., 2005; Onuma et al., 2007; Ejiri et al., 2008; Murata et al., 2007, 2008b; Sato et al., 2007; Imura et al., 2011; Tanigawa et al., 2013), but there seem to have been no outbreaks of the diseases among the wild birds in Japan . Among them, although non-positive identification (only genus level) for the most protozoan taxa detected were given, the genus *Plasmodium* from *Lanius bucephalus* was identified with *P. gallianceum* (Murata et al., 2007, 2008b). At least in Japan, hosts of the genus Babesia have been restricted to mammals, but the piroplasma was obtained from *Spilornis cheela* in Yaeyama Archipelago, one of most southern parts of Japan (Murata et al., 2008a).

Certain free living protozoans such as diatoms (Phylum Bacillariophyta: genera *Nitzschia* and *Pseudonitzschia*) and dinoflagellates (Phylum Dinoflagellata: genera *Alexandrium* and *Gymnodinium*) produce neurotoxins such as domoic acid, saxitoxin, and brevetoxin, that can affect sea birds (Asakawa et al., 2002). Most reports of mortality in birds have been die-offs occurring in conjunction with protozoan blooms in the United States (Friend and Franson 1999). Although no such mortalities have been reported yet in Japan, many protozoan blooms occur around Japan making it essential that we pay attention to such events because of their potential for causing mortality.

Conclusion

Disease outbreak prevention will need to be based not only on an understanding of avian ecology but also of agents (pathogens) epidemiology. And, in order for medical researchers to be able to prepare for future epidemic outbreaks amongst birds, they must perform epidemiological surveys of the probable agents mentioned above in wild bird populations both in Japan and its surrounding areas; international and inter-agency cooperation is, therefore, essential in disease prevention. Very careful attention should be paid to the possibility of infection by zoonotic agents during monitoring surveys.

Acknowledgments

The present chapter was supported in part by the priority research (2502-11E004) of Hokkaido Research Organization, and by the Supported Program for the Strategic Research Foundation at Private Universities (2013) of the Ministry of Education, Culture, Science and

Technology, Japan, respectively. We thank Mr. Y. Honma, Hokkaido Shinbun Co., Ltd., and Dr. T. Yoshino, Ministry of the Environment in Japan, for reading the manuscript and giving their kindly comments.

References

Asakawa, M. 2006. [(translated title) Intimate relationship between forensic veterinary medicine and zoo/wild animal medicine with special reference to mass mortality of wild birds in Japan]. *Zoo and Wildlife News*, Japan, (22): 46–53. (in Japanese)

Asakawa, M. 2010. Ecotourism with utilization of wild animals - Its impact on conservation medicine and risk assessment in Hokkaido, Japan. In: (Eds. Anton Krause and Erich Weir) *Ecotourism: Management, Development and Impact*, Nova Science Publishers, Inc., New York: 227-240.

Asakawa, M., Horikami, A., Wada, M., Aizawa, K., Watanabe, H., Yoshino, T., Okamoto, M., Onuma, M., Murata, K., Kuwana T. 2008. Post mortem examination of mass death carcasses of Short-tailed Shearwater (*Puffinus tenuirostris*) found in northern Hokkaido, Japan, with special reference to parasitological observation. *Jpn. J. Vet. Parasitol.*, 7 (1): 21. (in Japanese)

Asakawa, M. and Kitamura, K.2003. Outbreaks of infectious diseases recorded in zoo and aquariums in Japan with overview of references cited in Journal of Japanese Association of Zoological Gardens and Aquariums. *J. Rakuno Gakuen Univ., Nat. Sci.,* 28: 79-84. (in Japanese with English summary).

Asakawa, M., Nakade, T., Murata, S., Ohashi, K., Osa, Y. and Taniyama, H. 2013. Recent viral diseases of Japanese anatid with a fatal case of Marek's disease in an endangered species, white-fronted goose (*Anser albifrons*). *In:* (Hambrick, J. and Gammon, L. T. Eds.) *Ducks: Habitat, Behavior and Diseases*, Nova Science Publishers, Inc., USA: 37-48.

Asakawa, M., Nakamura, S. and Brazil, M. A. 2002. An overview of infectious and parasitic diseases in relation to the conservation biology of the Japanese avifauna. *J. Yamashina Inst. Ornithol.,* 34: 200-221.

Asaoka, Y., Yanai, T., Hirayama, H., Une, Y., Saito, E., Sakai, H., Goryo, M., Fukush,i H. and Masegi, T. 2004. Fatal necrotic enteritis associated with *Clostridium perfringens* in wild crows (*Corvus macrorhynchos*). *Avian Pathol.,* 33: 19-24.

Bui, V. N., Ogawa, H., Xininigen, K. K., Matsuo, K., Awad, S. S., Minoungou, G. L., Yoden, S., Haneda, H., Ngo, L. H., Tamaki, S., Yamamoto, Y., Nakamura, K., Saito, K., Watanabe, Y., Runstadler, J., Huettmann, F., Happ, G. M. and Imai, K. 2012. H4N8 subtype avian influenza virus isolated from shorebirds contains a unique PB1 gene and causes severe respiratory disease in mice. *Virology*. 426: 83-85.

Committee for Check-List of Japanese Birds of Ornithological Society of Japan (Ed.). 2012. *Check-List of Japanese Birds, 7th ed. Ornithological Society of Japan*, Tokyo, Japan: 438 pp.

Ejiri, H., Sato, Y., Sasaki, E., Sumiyama, D., Tsuda, Y., Sawabe, K., Matsui, S., Horie, S., Akatani, K., Takagi, M., Omori, S., Murata, K. and Yukawa, M. 2008. Detection of avian *Plasmodium* spp. DNA Sequences from mosquitoes captured in Minami Daito Island of Japan. *J. Vet. Med. Sci.*, 70: 1205-1210.

Friend, M. and Franson, J. C. (Eds.) 1999. *Field Manual of Wildlife Diseases*, USGS, Washington D. C.: 426 pp.

Fujii, K. and Osa, Y. 2013. Infectious diseases shared by wildlife and livestock-Epidemiological investigation of *Salmonell* as an example. *Jpn. J. Zoo. Wildl. Med.*, 18: 61-64. (in Japanese with English summary).

Fujimoto, Y., Ito, H., Shivakoti, S., Nakamori, J., Tsunekuni, R., Otsuki, K. and Ito, T. 2010. Avian influenza virus and paramyxovirus isolation from migratory waterfowl and shorebirds in San-in district of western Japan from 2001 to 2008. *J. Vet. Med. Sci.*, 72: 963-967.

Fukuda, M., Yanai, T., Sakai, H., Masegi, T., Mori, Y. and Ito, H. 2008. [(translated title) Suspicious case of *Mycobacterium genavense* infection to *Geopelia cuneata*]. The *Proceedings of 14th Annual Meeting of Japanese Society of Zoo and Wildlife Medicine,* Kobe University, Japan: 131. (in Japanese).

Fukui, D. 2013. Infectious diseases associated with relation between humans and wildlife-Consideration on wild bird mortality and bird-feeding. *Jpn. J. Zoo. Wildl. Med.*, 18: 41-48. (in Japanese with English summary).

Fukui, D., Bando, G., Yokota, H., Shibahara, T., Kadota, K., Asakawa, M. and Kosuge, M. 2006. [(translated title) Case report on mass mortality of sparrows due to Atoxoplasmosis with retrospective analysis for annual records carried to zoo animal hospital between 1999 and 2006]. *J. Hokkaido Vet. Med. Assoc.*, 50: 363. (in Japanese).

Fukui, D., Takahashi, K., Une, Y., Kato, Y., Izumiya, H., Asakawa, M. and Teraoka, H. 2009. [(translated title) Mass mortality of sparrows (*Passer montanus*) due to the Salmonella infectious in Kamikawa, Hokkaido, Japan]. *J. Hokkaido Vet. Med. Assoc.*, 53: 486. (in Japanese).

Fukuyama, M., Furuhata, K., Oonaka, K., Sakata, S., Hara, M., Kakuno, Y., Itoh, T., Kai, A., Obata, H. and Watanabe, T. 2003. Isolation and serotypes of Vero toxin-producing *Escherichia coli* (VTEC) from pigeons and crows. Kansenshogaku Zasshi., 77: 5-9. (in Japanese with English summary).

Hayakawa, D., Ikeda, H., Yoshizawa, M., Yamasaki, H. and Watabe, T. 2011. A case for the detection of multiple drug-resistant bacteria causing food poisoning in raptor on wildlife rescue. *The Proceedings of 17th Annual Meeting of Japanese Society of Zoo and Wildlife Medicine*, Tokyo University of Agriculture and Technology, Japan: 104. (in Japanese).

Hasegawa, Y. 2013. [(translated title) A human case of avian tuberculosis infected from a captive diamond dove (Geopelia cuneata)]. *The Proceedings of 198th Nihon-niaka-gakkai-Tohoku-chikureiki,* Tohoku University: page unknown. (in Japanese).

Hisada, Y., Saito, K. and Asakawa, M. 2004. Epidemiological survey of *Haemoproteus* sp. Found Blakiston's Owls (*Ketupa blakistoni blakistoni*) on Hokkaido Island, Japan. *Jpn. J. Zoo Wildl. Med.*, 9: 85-89. (in Japanese with English summary).

Honma, H., Suyama, Y., Watanabe, Y., Matsumoto, F. and Nakai, Y. 2011. Accurate analysis of prevalence of coccidiosis in individually identified wild cranes in inhabiting and migrating populations in Japan. *Environ. Microbiol.* 13: 2876-2887.

Imura, T., Suzuki, Y., Ejiri, H., Sato, Y., Ishida, K., Sumiyama, D., Murata, K. and Yukawa, M. 2011. Prevalence of avian haematozoa in wild birds in a high-altitude forest in Japan. *Vet. Parasitol.*, 183: 244-248.

Ishimoto, A., Yamanaka, M. and Araki, Y. 2007. The pathological survey of injured and diseased wild birds in Shiga Prefecture. *Jpn. J. Soc. Zoo Wildl. Med.*, 12: 87-95. (in Japanese with English summary).

Jahangir, A., Ruenphet, S., Ueda, S., Ueno, Y., Shoham, D., Shindo, J., Okamura, M., Nakamura, M. and Takehara, K. 2009. Avian influenza and Newcastle disease viruses from northern pintail in Japan: isolation, characterization and inter-annual comparisons during 2006-2008. *Virus Res.,* 143: 44-52.

Kawamori, F., Arita, Y., Nishio, T., Miwa, N., Masuda, T. and Akiyama, M. 2004. Prevalence of *Campylobacter* spp. and *Helicobacter* spp. in patients with enteritis, dogs, cats and wild birds and comparison of isolation methods. *J. Jpn. Vet. Med. Assoc.*, 57: 455-459. (in Japanese with English Summary).

Kitadai, N., Obi, T., Yamashita, S., Murase, T. and Takase, K. 2012. Antimicrobial susceptibility of *Escherichia coli* isolated from feces of wild cranes migrating to Kagoshima, Japan. *J. Vet. Med. Sci.*, 74: 395-397.

Kobayashi, H., Kanazaki, M., Hata, E. and Kubo, M. 2009. Prevalence and characteristics of eae- and stx-positive strains of *Escherichia coli* from wild birds in the immediate environment of Tokyo Bay. *Appl Environ Microbiol.*, 75: 292-295.

Kobayashi, H., Kanazaki, M., Shimazu, Y., Nakajima, H., Khatun, M. M., and Kubo, M. 2007. *Salmonella* Isolates from cloacal swabs and footpads of wild birds in the immediate environment of Tokyo Bay. *J Vet Med Sci.,* 69: 309-311.

Kubo, M. 2006. Sarcocysts observed in wild birds. *Bull. Nat. Inst. Anim. Heal.*, (112) : 19-24. (in Japanese).

Kubo, M., Tanimura, N. and Goto, Y.2005. Pathology of wild birds. *Bull. Nat. Inst. Anim. Heal.*, (111) : 9-20. (in Japanese).

Kuroki, T., Izumiyama, S., Yagita, K., Miyake, Y. and Endo, T. 2006. [(translated title) *Cryptosporidium* infection in wild birds]. *J. Vet. Epidemiol.*, 10: 13. (in Japanese).

Mase, M., Inoue, T. and Imada, T. 2009. Genotyping of Newcastle disease viruses isolated from 2001 to 2007 in Japan. *J. Vet. Med. Sci.*, 71: 1101-1104.

Matsubayashi, M., Takami, K., Abe, N., Kimata, I., Tani, H., Sasai, K. and Baba, E. 2005. Molecular characterization of crane Coccidia, *Eimeria gruis* and *E. reichenowi*, found in feces of migratory cranes. *Parasitol. Res.*, 97: 80-33.

Murao, T., Omata, Y., Kano, R., Murata, S., Okada, T., Konnai, S., Asakawa, M., Ohashi, K. and Onuma, M. 2008. Serological survey of *Toxoplasma gondii* in wild waterfowl in Chukotka, Kamchatka, Russia and Hokkaido, Japan. *J. Parasitol.,* 94: 830-833.

Murata, K., Itoh, K., Sasaki, E., Sato, Y., Kinjo, T., Amano, Y. and Nagamine T. 2008a. Avian piroplasm *Babesia* sp. isolated from Crested Serpent Eagles (*Spilornis cheela*) in Yaeyama Archipelago. *Jpn. J. Zoo Wildl. Med.*, 13: 29-33. (in Japanese with English Summary).

Murata, K., Nii, R., Yui, S., Sasaki, E., Ishikawa, S., Sato, Y., Matsui, S., Horie, S, Akatani, K.Takagi, M., Sawabe, K. and Tsuda, Y. 2008b. Avian haemosporidian parasites infection in wild birds inhabiting Minami-Daito Island of the Northwest Pacific, Japan. *J. Vet. Med. Sci.*, 70: 501-503.

Murata, K., Tamada, A., Ichikawa, Y., Hagihara, M., Sato, Y., Nakamura, H., Nakamura, M., Sakanakura, T., Asakawa, M. 2007. Geographical distribution and seasonality of the prevalence of *Leucocytozoon lovati* in Japanese rock ptarmigans (*Lagopus mutus japonicus*) found in the alpine regions of Japan. *J. Vet. Med. Sci.*, 69: 171-176.

Nagata, H. and Sodhi, N. 2003. Low prevalence of blood parasites in five *Sylviidae species* in Japan. *Ornithol. Sci.*, 2: 73-74.

Nakamura, S., Endoh, D., Sato, J. and Asakawa, M. 2004. Preliminary survey of Marek's Disease Virus of wild geese kept in Niigata Prefectural Bird Protection Center, Japan. *J. Rakuno Gakuen Univ., Nat., Sci.,* 28: 163-165.

Nakayama, M. and Isobe, T. 2004. Detection of pathogens from wild birds in Niigata Prefecture. *J. Jpn. Soc. Poultry Dis.*, 40: 100-104. (in Japanese English summary).

Neagari, Y. 2013. Challenges for the investigation of causes of wildlife mortality. *Jpn. J. Zoo Wildl. Med.*, 18: 53-59. (in Japanese with English summary).

Nishimori-Ohtsuka, E., Sanada, N., Sanada, Y., Takeuchi, Y. and Asakawa, M. 2009. Epidemiological surveys on gastro-intestinal protozoans in captive birds with case report of clinical examination. *Proc. Jpn. Assoc. Clin. Avian Med.*, (12): 9-13 (in Japanese with English summary).

Ohnishi, Y., Izawa, K., Ohori, K., Edagawa, A., Kimura, A., Chahota, R. and Fukushi, H. 2006. Epidemiological studies of *Chlamydophila psittaci* in pet or captive birds. The *Proceedings Joint International Symposium of the 12th Annual Meeting of Japanese Society of Zoo and Wildlife Medicine and the 21st Century COE Program of Gifu University, Japan*: 158. (in Japanese with English summary).

Okamatsu, M., Tanaka, T., Yamamoto, N., Sakoda, Y., Sasaki, T., Tsuda, Y., Isoda, N., Kokumai, N., Takada, A., Umemura, T. and Kida, H. 2010. Antigenic, genetic, and pathogenic characterization of H5N1 highly pathogenic avian influenza viruses isolated from dead whooper swans (*Cygnus cygnus*) found in northern Japan in 2008. *Virus Genes.*, 41: 351-357.

Onuma, M., Kuwana, T. and Asakawa, M. : [(translated title) Risk assessment of extinction due to an outbreak of West Nile disease among free ranging Japanese Crane (*Grus japonensis*) in Hokkaido, Japan, with usage of the software "Vortex"]. *J Hokkaido Vet Med Assoc*, 54: 311-312, 2010. (in Japanese)

Onuma, M., Ohshima, Y., Hisada, Y., Okamoto, M., Uebayashi, A., Shimura, R., Saito, K., Murata, K., Kuwana, T. and Asakawa, M., 2007. Epidemiological and pathological investigations of protozoan patrasites for endangered avian species in Japan. *Jpn. J. Parasitol.*, 6 (1): 27. (in Japanese).

Saito, K., Komada, A., Yamaguchi, T., Gotoh, Y., Sakai, H., Fukushi, H., Masegi, T. and Yanai, T. 2009. Avian poxvirus infection in a white-tailed sea eagle (*Haliaeetus*Avian *albicilla*) in Japan. *Avia. Pathol.*, 38: 485-489.

Saito, M., Ito, T., Amano, Y., Takara, J., Nakata, K., Tamanaha, S., Osa, Y., Yoshino, T., Onuma, M., Kuwana, T., Ogura, G., Tadano, M., Endoh, D. and Asakawa, M. 2011. Trials for risk assessment of Japanese encephalitis based on serologic survey of wild birds and animals. In: (Ružek, D. Ed.) *Flavivirus Encephalosis*, InTech, Croatia: 427-438.

Saito, M., Osa, Y. and Asakawa, M. 2009. Antibodies to flaviviruses in wild ducks captured in Hokkaido, Japan: Risk assessment of invasive flaviviruses. *Vec.-Born. Zoon. Dis.,* 9: 253-258.

Sarashina, T., Uzuka, Y., Tanabe, S., Oku, Y., Watanabe, Y., Kurosawa, N. and Nishimura, M. 2006. Survey of coccidial oocysts and parasite eggs in feces of free-ranging *Grus japonensis*. *J. Vet. Med. Sci.,* 68: 873-875.

Sato, A., Murata, K., Ikebe, Y., Kawai, N., Koizumi, J. and Asakawa, M. 2004. A case report on the parasitological survey performed in several zoos on Honshu I., Japan. *The Proceedings of 10th Annual Meeting of Japanese Society of Zoo and Wildlife Medicine,* University of Tokyo, Japan: 46. (in Japanese).

Sato, Y. 2011. [(translated title) Infectious diseases due to a avian protozoan parasites and their conservation medical view point.] *J. Vet. Med.*, Tokyo, 64: 15-20. (in Japanese).

Sato, Y., Hagiwara, M., Yanaguchi, T., Yukawa, M. and Murata, K. 2007. Phylogenetic comparison of *Leucocytozoon* spp. from wild birds of Japan. *J. Vet. Med. Sci*., .69: 55-59.

Shiibashi, T., Ishihara, S., Hagiwara, M., Nagamine, T., Nogami, S. and Murata, K. 2005. Parasitological investigation on the coccidium infection in Okinawa rail (*Gallirallus okinawae*). *The Proceedings of 11th Annual Meeting of Japanese Society of Zoo and Wildlife Medicine,* Obihiro University of Agriculture and Veterinary Medicine, Japan: 104. (in Japanese).

Sugiyama, M., Abe, M., Ito, N., Minamoto, N., Asano, M., Tsubota, T., Mizoguchi, T., Saita, E., Watanabe, G. and Taya, K. 2006. Seroepidemiological survey on group A rotaviruses in wild animals in Japan. *The Proceedings Joint International Symposium of the 12th Annual Meeting of Japanese Society of Zoo and Wildlife Medicine and the 21st Century COE Program of Gifu University,* Japan: 158. (in Japanese with English summary).

Suzuoka, N., Sugiyama, M., Ito, N. and Minamoto, N. 2002. Serological survey for rotavirus infection in wildlife. *The Proceedings of 8th Annual Meeting of Japanese Society of Zoo and Wildlife Medicine*, University of Tokyo, Japan: 81. (in Japanese).

Tanaka, T., Tanoue, G., Yamasaki, M., Takashima, I., Sakoda, Y., Ochiai, K. and Umemura, T. 2008. Chemical deicer poisoning was suspected as a cause of the 2005-2006 wintertime mortality of small wild birds in Hokkaido. *J. Vet. Med. Sci*., 70: 607-610.

Tanigawa, M., Sato, Y., Ejiri, H., Imura, T., Chiba, R., Yamamoto, H., Kawaguchi, M., Tsuda, Y., Murata, K. and Yukawa, M. 2013. Molecular identification of avian haemosporidia in wild birds and mosquitoes on Tsushima Island, Japan. *J. Vet. Med.* Sci., 75: 319-326.

Uchida Y., Suzuki, Y., Shirakura, M., Kawaguchi, A., Nobusawa, E., Tanikawa, T., Hikono, H., Takemae, N., Mase, M., Hanehira, K., Hayashi, T., Tagawa, Y., Tashiro, M. and Saito, T. 2012. Genetics and infectivity of H5N1 highly pathogenic avian influenza viruses isolated from chickens and wild birds in Japan during 2010-11. *J. Virusres* 09.004: 109-117.

Ueda, H., Adachi, T., Shimura, R., Mizuo, A., Nippashi, K., Ohshima, Y., Takita, Y., Sasaki, K. and Asakawa, M. 2008. [(Translated title) Fecal examination (parasitology) for captive birds and mammals in zoological gardens performed by WAMC in 2007].*The Proceedings of 14th Annual Meeting of Japanese Society of Zoo and Wildlife Medicine,* Kobe University: 101. (in Japanese).

Ueda, M., Une, Y., Yamamoto, Y., Matsui, K., Matsumoto, R., Mizutani, N. and Soumura, H. 2003. Suspected atoxoplasmosis in a Starling. *The Proceedings of 9th Annual Meeting of Japanese Society of Zoo and Wildlife Medicine,* University of the Ryukyus, Japan: 103. (in Japanese).

Watanabe, M., Alexandre, T. O., Shigeta, Y., Iwata, T., Horisaka, T., Horikita, T. and Hayashidani, H. 2002. Prevalence of *Vibrio* spp. in migratory aquatic birds in Japan. *J. Vet. Epidemiol.*, 6 (2): 77-83. (in Japanese with English summary) .

Watanabe, Y., Matsumoto, F. and Koga, K. 2003. A survey of the coccidian infection of wild Japanese Cranes *Grus japonensis* in Hokkaido. *J. Yamashina Inst. Ornithol.*, 35: 55-60. (in Japanese with Emglish summary).

Watanabe, Y., Sakai, J., Saito, K., Kanazawa, K. and Yanai, T. 2010. [(translated title) A case report of avipoxvirus infection in a white-tailed sea eagle permanent resident individual]. *J. Hokkaido Vet. Med. Assoc.*, 54: 129. (in Japanese).

Yamaguchi, T., Kitahara, M., Sakanakura, T., Murata, K., Simizu, H., Masegi, T.,Yanai, T., Sakai, H., Hirai, K. and Fukushi, H. 2003. Isolation of staphylococci living and captive rock ptarmigans. *The Proceedings of 9th Annual Meeting of Japanese Society of Zoo and Wildlife Medicine*, University of the Ryukyus, Japan: 103. (in Japanese).

Yamaguchi, T., Nakamori, S., Shimada, A., Fefuki, T., Ito, K, Ozaki, K. and Ito, T. 2008. [(translated title) Avipox in *Emberiza schoeniclus*]. *The Proceedings of 14th Annual Meeting of Japanese Society of Zoo and Wildlife Medicine,* Kobe University, Japan: 117. (in Japanese).

Yamamoto, N., Sakoda, Y., Motoshima, M., Yoshino, F., Soda, K., Okamatsu, M. and Kida, H. 2011. Characterization of a non-pathogenic H5N1 influenza virus isolated from a migratory duck flying from Siberia in Hokkaido, Japan, in October 2009. *Virol J.* 8: 65.

Yoshino, T. and Asakawa, M. 2012. [(translated title) An overview of postmortem examinations for wild water birds at Miyajima-numa.], In (Ushiyama, K Ed.) *Minnade-magan-wo-kazoerukai-25shunen-kinenshi*, Bibai-shi, Japan: 25-27. (in Japanese).

In: Birds
Editors: Lucas Ruiz and Franco Iglesias

ISBN: 978-1-62948-104-3

Chapter 6

Ticks on Brazilian Birds: Overview

Hermes Ribeiro Luz and João Luiz Horacio Faccini
Departamento de Parasitologia Animal, Instituto de Veterinária,
Universidade Federal Rural do Rio de Janeiro, Seropédica, RJ, Brazil

Abstract

Brazil has one of the richest avifauna in the world. There are 1,832 species distributed in 31 orders and 98 families. The Amazon, Atlantic Forest and Cerrado biomes have the richest diversity of birds in the country, but the most endemic bird species are found in the Atlantic Forest. Several species of migratory birds are also found in this country, originating from North and Central America and some countries in South America. Therefore, data from studies focusing parasitism by ticks and the pathogens they transmit are of great importance to animal and human welfare. To date, 6,201 ticks (4,270 larvae, 1,784 nymphs and 147 adults) have been collected. Passeriformes was the most parasitized order (5,964 ticks). The Thamnophilidae shows the highest numbers of parasitized species 25 (14%) out of 174 occurring in Brazil, followed by Tyrannidae (22/13.16%), Furnariidae (18/137. 17%) and Thraupidae (17/99. 17%). A total of 21 species of ticks belonging to four genera: *Amblyomma*, *Rhipicephalus*, *Ixodes* and *Haemaphysalis* have been identified. The most abundant tick species are *Amblyomma longirostre* and *Amblyomma nodosum*. To date only bacteria from the genus *Rickettsia* are isolated from ticks associated with wild birds: from *Amblyomma longirostre, Rickettsia amblyommi* and *Rickettsia parkeri;* from *Amblyomma nodosum, R. parkeri;* from *Amblyomma geayi, R. amblyommi;* from *Amblyomma parkeri, R. parkeri,* from *Amblyomma calcaratum, R. parkeri* and from *Amblyomma sp., R. parkeri.*

Keywords: Aves, Acari, Ixodidae, *Rickettsia*, Brazil

Introduction

The importance of birds to maintain biodiversity and ecological balance of nature is notorious. However, birds can spread several diseases such as avian influenza (bird flu), psittacosis and cryptosporidiosis, and transport over long distances the arthropod vectors of several pathogens which infect humans and domestic and companion animals (Hoogstraal, 1961; Santos-Silva et al., 2006; Elfving et al., 2010; Pacheco et al., 2012).

Ticks have a rather restricted movement and are mainly dispersed by birds in the stages of larvae and nymphs (immatures), although there are also reports on dispersion of adults (Barros-Battesti et al., 2003; Santos-Silva et al., 2006; Labruna et al., 2007; Szabó et al., 2008; Jordan et al., 2009; Ogrzewalska et al., 2009; Tolesano-Pascoliet al., 2010; Ogrzewalska et al., 2011; Luz et al., 2012; Pacheco et al., 2012; Pascoal et al., 2012; Sanches et al., 2013). Recent published papers have shown that birds serve as hosts for one or more life cycle stages of the same species or different species of ticks (Hubalek, 2004; Santos-Silva et al., 2006; Ogrzewalska et al., 2008; Ioannou et al., 2009; Jordan et al., 2009).

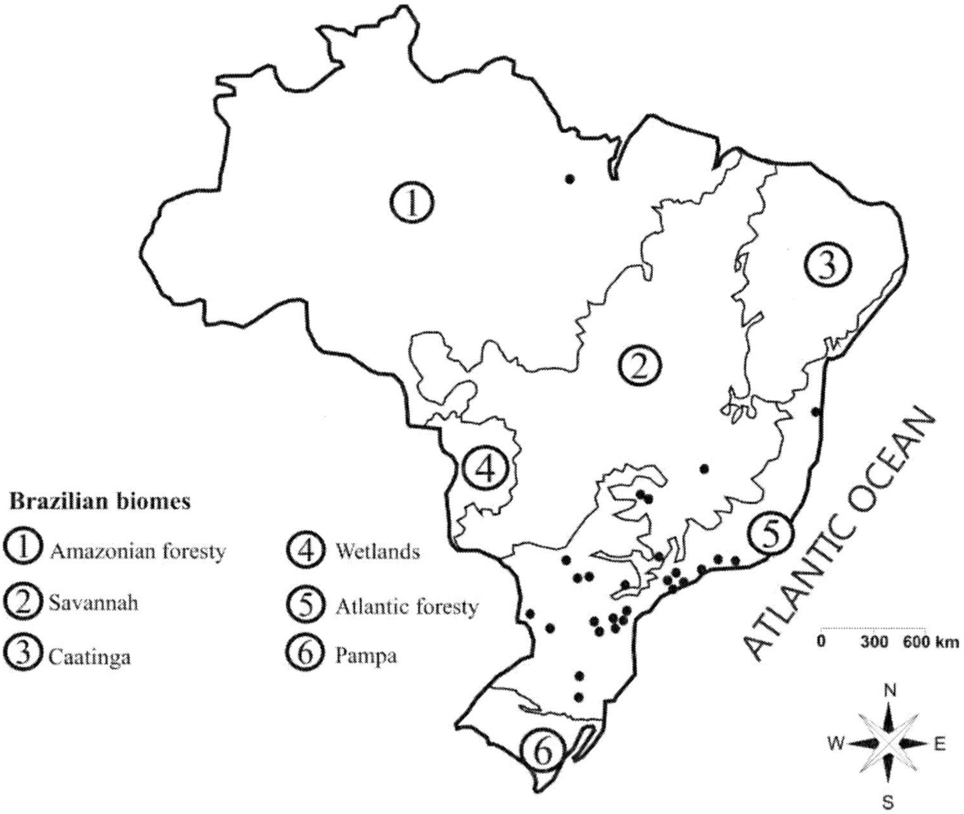

Figure 1. Distribution of main studies (black dots) of parasitism by ticks on birds in Brazil.

Emergence or reemergence of tick-borne diseases is of concern worldwide. In this scenario, a detailed knowledge on the relationship between ticks and wild birds can further assist to understand the dispersion of ticks through several regions of the planet, and consequently the spread of tick-borne diseases (Ogden et al., 2008; Hasle et al., 2009; Graham et al., 2010). The most important group of birds involved in the dispersion of ticks worldwide is the migratory birds which can introduce a given species of tick into a free-tick area, as is the case of Norway (Hasle et al., 2009). These birds can carry ticks either for long (between countries or continents) or short distances within the same country (Scott et al., 2001, Graham et al., 2010). In North America, the tick *Ixodes scapularis* Say is dispersed mainly by wild birds which are also incriminated as a possible reservoirs for *Borrelia burgdorferi* (causative agent of Lyme Disease) and *Anaplasma phagocytophilum* (causative agent of Human Granulocytic Anaplasmosis) (Daniels et al., 2002). However, it is the group which still needs to be better investigated.

Research on tick-borne diseases spread by birds is still scarce in Brazil, thus there is an urgent need to encourage research focusing on this issue.

The chapter is aimed to overview the known relations between birds and ticks in Brazil. Although such studies have advanced in the last 10 years, there are still gaps in knowledge about the relationship between tick and birds in this country. Currently, most of the papers cover surveys carried out in three of the six biomes defined for Brazil: Atlantic Forest, Amazon Forest and Cerrado (Figure 1). For the remaining three biomes, more detailed investigations are needed.

Ticks

Ticks are arthropods classified in the subclass Acari and suborder Ixodida. The suborder comprises three families: Argasidae (soft ticks), Ixodidae (hard ticks) and Nuttalliellidae. The first two families (Ixodidae and Argasidae) are the most diverse and widely distributed over several environments and a variety of hosts (Barros-Battesti, 2006). According to Guglielmone et al., (2010), ticks worldwide comprises about 896 species, 700 species in Ixodidae, 195 in Argasidae and one in Nuttalliellidae. There are at least 64 species reported for Brazil with 44 belonging to the Ixodidae, included in five genera: *Amblyomma* (30 species), *Ixodes* (8 species), *Haemaphysalis* (3 species), *Rhipicephalus* (2 species) and *Dermacentor* (1specie) and 20 species belonging to Argasidae included in four genera: *Argas* (1 species), *Antricola* (3 species), *Ornithodoros* (15 species) and *Nothoaspis* (1 species) (Dantas-Torres et al., 2009).

Relationship between Ticks and Wild Birds in Brazil

Ticks have variable selectivity for their hosts with some species only feeding on related groups of hosts while others are lesser selective. In general, the immature stages of two or three host ticks feed on small animals, while adults have preference for medium-sized or large hosts (Faccini and Barros - Battesti, 2006). Among small animals, the wild birds are hosts to a variety of ticks of the genera *Ixodes* (Arzua et al., 2003; Barros-Battesti et al., 2003; Arzua et al., 2005), *Rhipicephalus* (Szabó et al., 2008; Luz et al., 2012), *Haemaphysalis* (Arzua et al., 2005; Ogrzewalska et al., 2012) and *Amblyomma* (Labruna et al., 2007; Ogrzewalska et al., 2009; Ogrzewalska et al., 2010; Tolesano-Pascoli et al., 2010; Ogrzewalska et al., 2011; Luz et al., 2012; Ogrzewalska et al., 2012; Ogrzewalska et al., 2013; Pacheco et al., 2012; Pascoal et al., 2012; Santolin et al., 2012; Amaral et al., 2013; Sanches et al., 2013), the latter being the main genus parasitizing birds in Brazil. Species of *Amblyomma,* mainly larvae and nymphs, are commonly found parasitizing several wild birds taxonomically and ecologically distinct such as large ground birds as *Cariama cristata* (Linnaeus) (Red-legged Seriema) to small nectarivorous *Coereba flaveola* (Linnaeus) (Bananaquit).

Regarding the identification of ticks, morphology is routinely used for species identification, sometimes supported by molecular techniques. For the genus *Amblyomma* there are taxonomic keys for nymphs (Martins et al., 2010) and adults (Aragão and Fonseca, 1961; Onofrio et al., 2006). However, there is no taxonomic key for larvae because small numbers of larvae has been identified at species level to date, either by morphology, molecular methods or both. For the *Ixodes* there is a taxonomic key only for adults (Onofrio et al., 2009). The absence of keys is a limiting factor in specific identification and consequently in ecological research. An alternative to this problem, though laborious and time consuming, is to rear individual larvae to the nymph stage and / or adults for correct identification (Labruna et al., 2007).

Brazil has approximately 1,832 species of birds distributed in 98 families and 31 orders (CBRO, 2011). The order Passeriformes is the most diverse and one with the largest number of species, about 55% of the Brazilian avifauna. Brazilian birds are distributed throughout the country, from the pampa biome of southern state of Rio Grande do Sul (RS) to the state of Amazon (AM) in the far north. Among the six biomes defined for the country (Amazonian Forest, Atlantic Forest, Caatinga, Cerrado, Pantanal and Pampa) the first two have the highest number of bird species and high levels of endemism (Marini and Garcia, 2005). According to Sick (1997), 92% of Brazilian birds species are endemic and just over 8% are migratory.

To date, 39.8% (39/98) families of birds are listed as hosts of ticks of the Ixodidae in Brazil, being the passerines the most cited group (Santos-Silva et al., 2006; Labruna et al., 2007; Ogrzewalska et al., 2009; Luz et al., 2012; Pacheco et al., 2012; Ogrzewalska et al., 2013; Sanches et al., 2013). There are also some reports of ticks parasitizing non passerine birds such as Columbiformes, Columbidae (Arzua et al., 2005; Labruna et al., 2007; Luz et al., 2012; Santolin et al., 2012), Apodiformes, Trochilidae (Barros - Battesti et al., 2003; Arzua et al., 2005; Luz et al., 2012) and Strigiformes, Strigidae (Arzua et al., 2005; Ogrzewalska et al., 2009). There are no reports of argasids parasitizing wild birds in Brazil. *A. miniatus* is the only proven species associated with domestic chicken. *Ornithodoros capensis* which belongs to the group "*capensis*", associated with marine birds from Neotropical coastal areas and islands (Guglielmone et al., 2003), is also included in Brazilian tick list by Dantas-Torres et al. (2012) without definition of its bird hosts.

Overall, approximately 1,288 birds distributed in 200 species and 15 orders have been described as parasitized by ticks. The passerines, as mentioned before, is the one with the highest number of parasitized species (163/198, 82%) included in (21/39, 53.8%) families. Data collected from selected papers show that both passerines and non passerines are infested with approximately 6201 specimens of ticks accounting for 22 species, included in four genera (*Amblyomma*, *Ixodes*, *Haemaphysalis* and *Rhipicephalus*). Overall, the larvae (4,270, 68.8%) and nymph (1,784, 28.7%) are the most frequent stages and the adult stage (147, 2.3%) less common (Table 1). Details of tick – bird relationships are presented in the following paragraphs.

Table 1. Number of tick stages (ixodidae) associated with passerine and non passerine birds in Brazil

	Ticks	Tick stages			
		Larvae	Nymphs	Adults	Total
 PASSERINES	Number of ticks	4,160	1,689	115	5,964
	Prevalence (%)	70	28	2	
 NON PASSERINES	Number of ticks	110	95	32	237
	Prevalence (%)	46	40	13	
Total		4,270	1,784	147	6,201

Relationship between Ticks and Passerine Birds in Brazil

Table 3 and 4 shows details of parasitism by species of *Amblyomma* and *Ixodes* in passerine birds. The passerines is a group of birds varying in size and color, from the Double-collared Seedeater (*Sporophila caerulescens*) which measures 12 cm to Amazonian Umbrellabird (*Cephalopterus ornatus*) which measures 48 cm (Sick, 1997; CBRO, 2011). This group contains 1023 species distributed in 33 families and 928 species (CBRO, 2011). Passerines also contains the majority of endemic species in the country, with 20% and 18% inhabiting the Atlantic Forest and Amazon Rainforest biomes, respectively (Marini and Garcia, 2005). Parasitism by ticks is recorded in approximately 1205 specimens of birds, comprising 72% (21/33) families and 16% (163/1023) species. The Thamnophilidae shows the highest numbers of parasitized species 25 (14%) out of 174 occurring in Brazil, followed by Tyrannidae (22/13, 16%), Furnariidae (18/137, 17%) and Thraupidae (17/99, 17%) (Table 2). Also, the Thamnophilidae has the highest number of infested individuals – abundance (Figure 2). It is worth of noting that only record of a male of *Rhipicephalus sanguineus* (Latreille) in *C. flaveola* (Bananaquit), the only specie included in the Coerebidae (Szabó et al., 2008). To date, parasitized passerines yielded approximately 5964 specimens of ticks, classified as larvae (4,160, 70%), nymphs (1,689, 28%) and adults (115, 2%) and accounting for 21 species included in four genera: *Amblyomma*, *Ixodes*, *Haemaphysalis* and *Rhipicephalus*. The genus *Amblyomma* is the largest genus bearing 15 (71.4%) species: *Amblyomma longirostre* (Koch), *Amblyomma nodosum* Neumann, *Amblyomma aureolatum* Pallas, *Amblyomma calcaratum* Neumann, *Amblyomma ovale* Koch, *Amblyomma coelebs*

Neumann, *Amblyomma parkeri* Fonseca & Aragão, *Amblyomma cajennense* Fabricius, *Amblyomma naponense* Packard, *Amblyomma humerale* Koch, *Amblyomma geayi* Neumann, *Amblyomma parvum* Aragão, *Amblyomma tigrinum* Koch, *Amblyomma dubitatum* Neumann, *Amblyomma pseudoconcolor* Aragão. *A. longirostre*, *A. nodosum* and *A. aureolatum* are the three commonly reported tick species associated with passerine birds (Arzua et al., 2005; Santos-Silva, et al., 2006; Labruna et al., 2007; Ogrzewalska et al., 2008; 2009; 2010; 2011; 2012; Luz et al., 2012; Pacheco et al., 2012; Sanches et al., 2013) .

Table 2. The current associations of passerine birds and tick species in Brazil. NPS= number of passerine species by families according CRBO (2011), NP= number of parasitized species and P = prevalence (NP/NPSx100)

Families	NPS	NP	P(%)
Thamnophilidae	174	25	14
Tyrannidae	137	22	16
Furnariidae	103	18	17
Thraupidae	99	18	17
Dendrocolaptidae	43	15	35
Rhynchocyclidae	64	10	16
Emberizidae	65	10	15
Turdidae	17	7	41
Pipridae	36	6	17
Parulidae	23	6	26
Tityridae	29	5	17
Troglodytidae	17	4	24
Conopophagidae	6	3	50
Fringillidae	18	3	17
Incertae sedis	9	2	22
Cardinalidae	15	2	13
Corvidae	8	2	25
Scleruridae	7	2	29
Icteridae	41	1	2
Coerebidae	1	1	100
Vireonidae	16	1	6
Total	928	163	17

Table 3. The species of *Amblyomma* associated with wild birds in Brazil. Tick stages (LL= larva, NN= nymph, AA= adult). Tick species (Alo= *A. longorostre*, Aau= *A. aureolatum*, Ano= *A. nodosum*, Aca= *A. calcaratum*, Ahu=*A. humerale*, Apa= *A. pacae*, Aco= *A. coelebs*, Aov= *A. ovale*, Acj=*A. cajennense*, Ana= *A. naponense*, Age= *A. geayi*, Apv=*A. parvum*, Abr= *A. brasiliensis*, Atig= *A. tigrinum*, Aps= *A. pseudoconcolor*, Ama= *A. maculatum*, Asp= *Amblyomma* sp.

Birds	Ticks				References
PASSERINES	Species	LL	NN	AA	
Thamnophilidae Swainson, 1824					
Myrmotherula longipennis Pelzeln, 1868	Alo;	3			Ogrzewalska et al. (2010)
	Asp	3			Ogrzewalska et al. (2010)
Myrmotherula hauxwelli (Sclater, 1857)	Asp	1			Ogrzewalska et al. (2010)
Myrmornis torquata (Boddaert, 1783)	Asp	3			Ogrzewalska et al. (2010)
Myrmeciza squamosa Pelzeln, 1868	Aau		1		Ogrzewalska et al. (2012)
Formicivora melanogaster Pelzeln, 1868	Alo;		2		Luz et al. (2012)
	Asp	1			Luz et al. (2012)
Formicivora rufa (Wied, 1831)	Ano;		2		Ogrzewalska et al. (2009); Pascoal et al. (2012)
Thamnomanes caesius (Temminck, 1820)	Alo; Ano; Ahu	3	4		Ogrzewalska et al. (2010) ; Ogrzewalska et al. (2011)
	Asp	4	1		Ogrzewalska et al. (2010)
Thamnophilus caerulescens Vieillot, 1816	Alo; Ano; Aca; Apa; Aau	90	14	1	Arzua et al. (2005); Labruna et al. (2007) ; Ogrzewalska et al. (2008); Ogrzewalska et al. (2009);Tolesano-Pascoli et al. (2010); Ogrzewalska et al. (2012); Luz et al. (2012); Pacheco et al. (2012)
	Asp	29	11		Ogrzewalska et al. (2009); Ogrzewalska et al. (2012)
Thamnophilus torquatus Swainson, 1825	Ano		1		Luz et al. (2012)
	Asp	2			Luz et al. (2012)
Thamnophilus ruficapillus Vieillot, 1816	Aau	21			Arzua et al.(2005)
Thamnophilus pelzelni Hellmayr, 1924	Ano; Aco; Aov; Acj; Ana	2	220		Ogrzewalska et al. (2009)
		65	5		Ogrzewalska et al. (2009)
Dysithamnus mentalis (Temminck, 1823)	Alo; Ano; Aca; Acj	28	18		Labruna et al. (2007) ; Ogrzewalska et al. (2008); Ogrzewalska et al. (2009); Pacheco et al. (2012); Sanches et al. (2013)
	Asp	53	5		Labruna et al. (2007); Ogrzewalska et al. (2008); Ogrzewalska et al. (2009); Ogrzewalska et al. (2012)
Taraba major (Vieillot, 1816)	Alo;		1		Sanches et al. (2013)
Mackenziaena severa (Lichtenstein, 1823)	Alo;		1		Arzua et al. (2005)
Hylophylax naevius (Gmelin, 1789)	Alo; Aca; Ahu	4	2		Ogrzewalska et al. (2010)
	Asp	8	1		Ogrzewalska et al. (2010)

Table 3. (Continued)

Birds	Ticks			References
Hylophylax punctulatus (Des Murs, 1856)	Asp	1		Ogrzewalska et al. (2010)
Pyriglena leuconota (Spix, 1824)	Alo; Aco;Ahu	11	1	Ogrzewalska et al. (2010)
	Asp	4		Ogrzewalska et al. (2010)
Pyriglena leucoptera (Vieillot, 1818)	Alo; Aca; Aau	14	8	Labruna et al. (2007);Ogrzewalska et al. (2012); Pacheco et al. (2012); Sanches et al. (2012)
	Asp	83		Labruna et al. (2007); Ogrzewalska et al. (2012)
Myrmoborus myotherinus (Spix, 1825)	Alo;		1	Ogrzewalska et al. (2010)
Willisornis poecilinotus (Cabanis, 1847)	Alo; Ahu; Age	5	1	Ogrzewalska et al. (2010)
	Asp	22	1	Ogrzewalska et al. (2010)
Schistocichla leucostigma (Pelzeln, 1868)	Ahu	4		Ogrzewalska et al. (2010)
	Asp	5		Ogrzewalska et al. (2010)
Phlegopsis nigromaculata (d'O. & Lafresnaye, 1837)	Alo; Aco	2		Ogrzewalska et al. (2010)
	Asp	21		Ogrzewalska et al. (2010)
Drymophila squamata (Lichtenstein, 1823)	Ano		2	Ogrzewalska et al. (2011)
	Asp	1	1	Ogrzewalska et al. (2011)
Herpsilochmus longirostris Pelzeln, 1868	Acj		1	Tolesano-Pascoli et al. (2010)
Conopophagidae Sclater & Salvin, 1873				
Conopophaga lineata (Wied, 1831)	Alo; Ano; Aca; Aco; Apa; Acj; Aau	88	126	Arzua et al. (2005); Labruna et al. (2007);Ogrzewalska et al. (2008); Ogrzewalska et al. (2009); Pacheco et al. (2012); Ogrzewalska et al. (2012); Sanches et al. (2013)
	Asp	1106	35	Labruna et al. (2007); Ogrzewalska et al. (2008); Ogrzewalska et al. (2009); Ogrzewalska et al. (2012)
Conopophaga melanops (Vieillot, 1818)	Alo; Aca; Apa	2	2	Ogrzewalska et al. (2011)
	Asp	9	8	Labruna et al. (2007); Ogrzewalska et al. (2011)
Conopophaga aurita (Gmelin, 1789)	Asp		1	Ogrzewalska et al. (2010)
Scleruridae Swainson, 1827				
Sclerurus caudacutus (Vieillot, 1816)	Ahu	1		Ogrzewalska et al. (2010)
Sclerurus scansor (Ménétriès, 1835)	Asp	1		Ogrzewalska et al. (2012)
Dendrocolaptidae Gray, 1840				
Dendrocincla turdina (Lichtenstein, 1820)	Alo;	1	1	Arzua et al. (2005)
	Asp	7		Ogrzewalska et al. (2011)
Dendrocincla merula (Lichtenstein, 1829)	Alo;	6		Ogrzewalska et al. (2010)
	Asp	31		Ogrzewalska et al. (2010)
Deconychura longicauda (Pelzeln, 1868)	Alo;	4		Ogrzewalska et al. (2010)

Birds	Ticks			References
	Asp	39		Ogrzewalska et al. (2010)
Sittasomus griseicapillus (Vieillot, 1818)	Alo; Aca; Apa	8	4	Arzua et al. (2005);Labruna et a. (2007); Ogrzewalska et al. (2009); Ogrzewalska et al. (2012); Sanches et al. (2013)
	Asp	2		Ogrzewalska et al. (2012)
Certhiasomus stictolaemus (Pelzeln, 1868)	Alo;	1	1	Ogrzewalska et al. (2010)
	Asp	1		Ogrzewalska et al. (2010)
Glyphorynchus spirurus (Vieillot, 1819)	Alo; Age	10	1	Ogrzewalska et al. (2011); Ogrzewalska et al. (2010)
	Asp	17	2	Ogrzewalska et al. (2010)
Xiphorhynchus fuscus (Vieillot, 1818)	Alo;	3		Labruna et al. (2007)
	Asp	8		Labruna et al. (2007)
Xiphorhynchus elegans (Pelzeln, 1868)	Alo;	3	2	Ogrzewalska et al. (2010)
	Asp	1		Ogrzewalska et al. (2010)
Xiphorhynchus guttatus (Lichtenstein, 1820)	Alo;	1		Ogrzewalska et al. (2011)
Xiphocolaptes albicollis (Vieillot, 1818)	Aca		1	Pacheco et al. (2012)
Campylorhamphus falcularius (Vieillot, 1822)	Alo;	1		Labruna et al. (2007)
Lepidocolaptes squamatus (Lichtenstein, 1822)	Alo; Ano; Apa	16		Labruna et al. (2007); Ogrzewalska et al. (2012)
	Asp	2		Labruna et al. (2007); Ogrzewalska et al. (2012)
Dendrocolaptes hoffmannsi Hellmayr, 1909	Alo; Aca; Ahu; Age	46	6	Ogrzewalska et al. (2010)
	Asp	59		Ogrzewalska et al. (2010)
Dendrocolaptes platyrostris Spix, 1825	Ano		1	Ogrzewalska et al. (2009)
Hylexetastes brigidai Silva, Novaes & Oren, 1996	Alo; Age	7	2	Ogrzewalska et al. (2010)
	Asp	21		Ogrzewalska et al. (2010)
Furnariidae Gray, 1840				
Xenops minutus (Sparrman, 1788)	Alo; Ano	2	1	Ogrzewalska et al. (2011); Ogrzewalska et al. (2012)
	Asp	4	1	Labruna et a. (2007); Ogrzewalska et al. (2012); Ogrzewalska et al. (2011)
Automolus leucophthalmus (Wied, 1821)	Alo; Ano	37	5	Labruna et al. (2007); Ogrzewalska et al. (2012); Pacheco et al. (2012)
	Asp	28		Labruna et al. (2007); Sanches et al. (2013)
Automolus paraensis Hartert, 1902	Aau		3	Ogrzewalska et al. (2012);
	Asp	1		Ogrzewalska et al. (2010)
Anabazenops fuscus (Vieillot, 1816)	Alo;	1		Labruna et al. (2007)

Table 3. (Continued)

Birds	Ticks			References
Philydor atricapillus (Wied, 1821)	Alo; Apa	1	1	Arzua et al. (2005); Ogrzewalska et al. (2012)
	Asp	2		Ogrzewalska et al. (2012)
Philydor rufum (Vieillot, 1818)	Alo;		1	Labruna et al. (2007)
Anabacerthia amaurotis (Temminck, 1823)	Alo;		1	Ogrzewalska et al. (2012)
	Asp	1		Ogrzewalska et al. (2012)
Certhiaxis cinnamomeus (Gmelin, 1788)	Alo;		1	Luz et al. (2012)
Synallaxis ruficapilla Vieillot, 1819	Alo; Aau	1	9	Labruna et al. (2007); Ogrzewalska et al. (2012)
	Asp	7		Labruna et al. (2007); Ogrzewalska et al. (2012)
Synallaxis spixi Sclater, 1856	Alo;		1	Arzua et al. (2005)
Synallaxis cinerascens Temminck, 1823	Aau	1	4	Arzua et al. (2005)
Cichlocolaptes leucophrus (Jardine & Selby, 1830)	Acj	42		Arzua et al. (2005)
Furnarius rufus (Gmelin, 1788)	Aau	2	1	Arzua et al. (2005)
	Asp	2		Ogrzewalska et al. (2013)
Cranioleuca obsoleta (Reichenbach, 1853)	Aau	12	1	Arzua et al. (2005)
Cranioleuca pallida (Wied, 1831)	Aau	4		Arzua et al. (2005)
Pipridae Rafinesque, 1815				
Pipra fasciicauda Hellmayr, 1906	Alo;		1	Ogrzewalska et al. (2009)
	Asp		3	Ogrzewalska et al. (2009)
Manacus manacus (Linnaeus, 1766)	Alo; Ano; Aca; Apa	18	31	Arzua et al. (2005); Labruna et al. (2007); Ogrzewalska et al. (2011)
	Asp	59	8	Labruna et al. (2007); Ogrzewalska et al. (2011)
Dixiphia pipra (Linnaeus, 1758)	Alo; Ano; Aca; Age	3	7	Ogrzewalska et al. (2010); Ogrzewalska et al. (2011)
	Asp		1	Ogrzewalska et al. (2011)
Ilicura militaris (Shaw & Nodder, 1809)	Alo;	1	1	Labruna et al. (2007)
	Asp	6		Labruna et al. (2007)
Chiroxiphia caudata (Shaw & Nodder, 1793)	Alo; Ano; Aca; Apa	38	13	Arzua et al. (2005);Labruna et al. (2007); Ogrzewalska et al. (2012); Pacheco et al. (2012)
	Asp	89	1	Labruna et al. (2007); Ogrzewalska et al. (2008); Ogrzewalska et al. (2012)
Antilophia galeata (Lichtenstein, 1823)	Alo;	18	9	Sanches et al. (2013) (4); Tolesano-Pascoli et al. (2010) (38)
	Asp	8		Sanches et al. (2013)
Tityridae Gray, 1840				
Onychorhynchus coronatus (Statius Muller, 1776)	Alo;	3		Ogrzewalska et al. (2010)
	Asp	6		Ogrzewalska et al. (2010)

Birds	Ticks				References
Myiobius barbatus (Gmelin, 1789)	Alo;	4	4		Ogrzewalska et al. (2011); Ogrzewalska et al. (2010)
	Asp	2			Ogrzewalska et al. (2011)
Schiffornis virescens (Lafresnaye, 1838)	Alo;	14			Ogrzewalska et al. (2012)
	Asp	18			Ogrzewalska et al. (2012)
Schiffornis turdina (Wied, 1831)	Alo;	3			Ogrzewalska et al. (2010)
	Asp	14			Ogrzewalska et al. (2010)
Pachyramphus polychopterus (Vieillot, 1818)	Ano		1		Luz et al. (2012)
Incertae sedis					
Platyrinchus mystaceus Vieillot, 1818	Alo; Apa; Aau	22	10		Labruna et al. (2007); Ogrzewalska et al. (2008); Ogrzewalska et al. (2009); Ogrzewalska et al. (2012)
	Asp	26			Ogrzewalska et al. (2012)
Platyrinchus saturatus Salvin & Godman, 1882	Asp	1	1		Ogrzewalska et al. (2010)
Rhynchocyclidae Berlepsch, 1907					
Mionectes oleagineus (Lichtenstein, 1823)	Alo	2			Ogrzewalska et al. (2011)
	Asp	3			Ogrzewalska et al. (2011)
Mionectes macconnelli (Chubb, 1919)	Alo	4			Ogrzewalska et al. (2010)
	Asp	5			Ogrzewalska et al. (2010)
Mionectes rufiventris Cabanis, 1846	Alo	10			Labruna et al. (2007); Ogrzewalska et al. (2012)
	Asp	21			Labruna et al. (2007); Ogrzewalska et al. (2012)
Leptopogon amaurocephalus Tschudi, 1846	Alo	9	9		Labruna et al. (2007); Ogrzewalska et al. (2009) Tolesano-Pascoli et al. (2010); Ogrzewalska et al. (2012); Ogrzewalska et al. (2009); Ogrzewalska et al. (2012); Pacheco et al. (2012)
	Asp	2			Ogrzewalska et al. (2008)
Tolmomyias sulphurescens (Spix, 1825)	Alo	3	4	2	Ogrzewalska et al. (2012); Arzua et al. (2005); Sanches et al. (2013)
	Asp	1			Ogrzewalska et al. (2012)
Tolmomyias poliocephalus (Taczanowski, 1884)	Alo; Apa	2			Ogrzewalska et al. (2012)
Hemitriccus minor (Snethlage, 1907)	Alo	1			Ogrzewalska et al. (2010)
Hemitriccus nidipendulus (Wied, 1831)	Alo		2		Labruna et al. (2007)
Hemitriccus margaritaceiventer (d'O & L., 1837)	Alo; Ano	1	8		Ogrzewalska et al. (2009); Luz et al. (2012)
	Asp	1			Luz et al. (2012)
Phylloscartes ventralis (Temminck, 1824)	Asp	1			Labruna et al. (2007)
Tyrannidae Vigors, 1825					

Table 3. (Continued)

Birds	Ticks				References
Elaenia flavogaster (Thunberg, 1822)	Alo;		3		Luz et al. (2012)
	Asp	3			Luz et al. (2012); Ogrzewalska et al. (2012)
Elaenia mesoleuca (Deppe, 1830)	Alo; Ano		8		Luz et al. (2012)
Elaenia cristata Pelzeln, 1868	Alo; Ano		11	1	Luz et al. (2012); Pascoal et al. (2012)
	Asp	2	1		Luz et al. (2012); Pascoal et al. (2012)
Elaenia obscura (d'Orbigny & Lafresnaye, 1837)	Ano		1		Pascoal et al. (2012)
	Asp		1		Pascoal et al. (2012)
Elaenia chiriquensis Lawrence, 1865	Ano		1		Pascoal et al. (2012)
	Asp		3		Pascoal et al. (2012)
Myiopagis viridicata (Vieillot, 1817)	Alo;	1			Labruna et al. (2007)
	Asp	1			Labruna et al. (2007)
Attila rufus (Vieillot, 1819)	Alo;		1		Labruna et al. (2007)
Attila spadiceus (Gmelin, 1789)	Ahu	1			Ogrzewalska et al. (2010)
	Asp	3			Ogrzewalska et al. (2010)
Myiarchus tuberculifer (d'O. & L., 1837)	Alo;		3		Luz et al. (2012)
Myiarchus ferox (Gmelin, 1789)	Alo; Aov		10		Arzua et al. (2005); Luz et al. (2012)
			2		Pascoal et al. (2012)
Casiornis rufus (Vieillot, 1816)	Alo; Ano; Aca; Apv		15		Ogrzewalska et al. (2009); Luz et al. (2012); Pascoal et al. (2012); Sanches et al. (2013)
	Asp	1			Luz et al. (2012)
Pitangus sulphuratus (Linnaeus, 1766)	Alo; Acj; Apv		9		Luz et al. (2012); Santolin et al. (2012); Pacheco et al. (2012)
Myiodynastes maculatus (Statius Muller, 1776)	Alo;		2		Arzua et al. (2005)
Myiozetetes similis (Spix, 1825)	Alo;		3		Luz et al. (2012)
Tyrannus melancholicus Vieillot, 1819	Alo; Aov		11		Luz et al. (2012)
Tyrannus savana Vieillot, 1808	Alo; Apv		20		Luz et al. (2012)
	Asp	1			Luz et al. (2012)
Myiophobus fasciatus (Statius Muller, 1776)	Alo;	1			Labruna et al. (2007)
Cnemotriccus fuscatus (Wied, 1831)	Alo; Ano		6		Ogrzewalska et al. (2009)
Lathrotriccus euleri (Cabanis, 1868)	Alo;		15		Arzua et al. (2005); Luz et al. (2012); Sanches et al. (2013)
	Asp	7			Luz et al. (2012); Sanches et al. (2013)
Xolmis cinereus (Vieillot, 1816)	Alo;		6		Luz et al. (2012)
	Asp	4			Luz et al. (2012)
Xolmis velatus (Lichtenstein, 1823)	Alo; Apv	4			Luz et al. (2012)

Birds	Ticks				References
	Asp	1			Luz et al. (2012)
Cnemotriccus fuscatus (Wied, 1831)			5		Ogrzewalska et al. (2009)
Vireonidae Swainson, 1837					
Cyclarhis gujanensis (Gmelin, 1789)	Alo; Aov		8		Labruna et al. (2007); Pacheco et al. (2012)
Corvidae Leach, 1820					
Cyanocorax cristatellus (Temminck, 1823)	Alo;		20		Luz et al. (2012)
	Asp	15			Luz et al. (2012)
Cyanocorax chrysops (Vieillot, 1818)	Asp		1		Ogrzewalska et al. (2009)
Troglodytidae Swainson, 1831					
Cantorchilus leucotis (Lafresnaye, 1845)	Alo; Ano		4		Tolesano-Pascoli et al. (2010)
	Asp	2			Tolesano-Pascoli et al. (2010)
Troglodytes musculus Naumann,	Aov; Aau	186	21		Arzua et al. (2005); Luz et al. (2012)
Microcerculus marginatus (Sclater, 1855)		3			Ogrzewalska et al. (2010)
Turdidae Rafinesque, 1815					
Turdus rufiventris Vieillot, 1818	Alo; Aov;Aau	198	200	1	Arzua et al. (2005); Labruna et al. (2007);Luz et al. (2012) Ogrzewalska et al. (2010); Ogrzewalska et al. (2012); Amaral et al. (2013)
	Asp	15			Labruna et al. (2007); Ogrzewalska et al. (2012); Sanches et al. (2013)
Turdus leucomelas Vieillot, 1818	Alo; Ano; Aca; Aau	6	19		Tolesano-Pascoli et al. (2010);Ogrzewalska et al. (2011); Pacheco et al. (2012); Sanches et al. (2013)
	Asp	18	3		Ogrzewalska et al. (2011); Pascoal et al. (2012); Ogrzewalska et al. (2013); Sanches et al. (2013)
Turdus fumigatus Lichtenstein, 1823	Alo;		1		Labruna et al. (2007)
Turdus amaurochalinus Cabanis, 1850	Alo; Ano; Acj; Aau; Aco		58	1	Arzua et al. (2005); Santolin et al. (2012); Ogrzewalska et al. (2009); Ogrzewalska et al. (2012); Luz et al. (2012); Amaral et al. (2013)
	Asp		2		Pascoal et al. (2012)
Turdus subalaris (Seebohm, 1887)	Alo; Aov		2		Arzua et al. (2005); Ogrzewalska et al. (2009)
	Asp	1			Sanches et al. (2013)
Turdus albicollis Vieillot, 1818	Alo; Aca; Aov; Acj; Aau	23	40		Arzua et al. (2005); Storni et al. (2005); Labruna et al. (2007); Ogrzewalska et al. (2010); Ogrzewalska et al. (2012); Pacheco et al. (2012)
	Asp	8	1		Labruna et al. (2007); Ogrzewalska et al. (2010)
Turdus flavipes Vieillot, 1818	Alo				Aragão (1911)*
Thraupidae Cabanis, 1847					
Saltator maximus (Statius Muller, 1776)	Alo; Ano	2	7		Aragão (1911)*; Tolesano-Pascoli et al. (2010); Ogrzewalska et al. (2011)
Saltator similis d'Orbigny & Lafresnaye, 1837	Alo; Ano; Aca; Aau	12	25		Arzua et al. (2005); Labruna et al. (2007); Luz et al. (2012); Pacheco et al. (2012); Pascoal et al. (2012)

Table 3. (Continued)

Birds	Ticks			References
	Asp	4	6	Labruna et al. (2007); Pascoal et a. (2012)
Nemosia pileata (Boddaert, 1783)	Alo;		5	Luz et al. (2012)
Tachyphonus coronatus (Vieillot, 1822)	Alo; Aca; Acj; Aau	24	50	Arzua et al. (2005); Labruna et al. (2007); Teixeira et al. (2008); Ogrzewalska et al. (2012); Pacheco et al. (2012); Sanches et al. (2013)
	Asp	43	2	Labruna et al. (2007); Sanches et al. (2013); Ogrzewalska et al. (2012)
Ramphocelus bresilius (Linnaeus, 1766)	Alo; Acj	15	21	Labruna et al. (2007)
	Asp	5	7	Labruna et al. (2007)
Ramphocelus carbo (Pallas, 1764)	Alo; Ano; Apv; Aca		63	Luz et al. (2012); Santolin et al. (2012); Ogrzewalska et al. (2013)
	Asp	15		Luz et al. (2012)
anio surinamus (Linnaeus, 1766)	Alo; Ahu	2		Ogrzewalska et al. (2010)
	Asp	5		Ogrzewalska et al. (2010)
Lanio penicillatus (Spix, 1825)	Alo; Ano		11	Tolesano-Pascoli et al. (2010); Sanches et al. (2013)
Lanio melanops (Vieillot, 1818)	Alo; Aca; Apa; Aau	26	33	Arzua et al. (2005); Labruna et al. (2007) ; Ogrzewalska et al. (2009); Ogrzewalska et al. (2011); Ogrzewalska et al. (2012);Pacheco et al. (2012); Sanches et al. (2013);
	Asp	104		Labruna et al. (2007); Ogrzewalska et al. (2008); Ogrzewalska et al. (2012); Sanches et al. (2013)
Tangara seledon (Statius Muller, 1776)	Alo;		5	Labruna et al. (2007); Arzua et al. (2005)
	Asp		2	Labruna et al. (2007)
Tangara sayaca (Linnaeus, 1766)	Alo;	1		Labruna et al. (2007)
	Asp	3		Labruna et al. (2007)
Tangara palmarum (Wied, 1823)	Alo;		19	Luz et al. (2012)
Tangara cayana (Linnaeus, 1766)	Alo; Apv		13	Luz et al. (2012)
	Asp	2	1	Luz et al. (2012); Pascoal et a. (2012)
Tersina viridis (Illiger, 1811)	Alo;		1	Arzua et al. (2005)
Pipraeidea melanonota (Vieillot, 1819)	Alo;		2	Ogrzewalska et al. (2012); Arzua et al. (2005)
Dacnis cayana (Linnaeus, 1766)	Alo;		11	Labruna et al. (2007); Luz et al. (2012)
Pyrrhocoma ruficeps (Strickland, 1844)	Acj	3		Arzua et al. (2005)
Paroaria capitata (d'Orbigny & Lafresnaye, 1837)	Asp	1		Ogrzewalska et al. (2013)
Emberizidae Vigors, 1825				
Arremon flavirostris Swainson, 1838	Ano; Aca; Aov; Acj	5	1	Ogrzewalska et al. (2009); Tolesano-Pascoli et al. (2010); Sanches et al. (2013)
	Asp	584	2	Labruna et al. (2007); Ogrzewalska et al. (2009)
Haplospiza unicolor Cabanis, 1851	Alo;		1	Ogrzewalska et al. (2012)
	Asp	1		Ogrzewalska et al. (2012)
Sporophila lineola (Linnaeus, 1758)	Alo;		2	Luz et al. (2012)

Birds	Ticks				References
Sporophila nigricollis (Vieillot, 1823)	Alo; Aov; Apv		13	1	Luz et al. (2012)
Sporophila caerulescens (Vieillot, 1823)	Alo; Ano; Apv		7		Luz et al. (2012)
	Asp	3			Luz et al. (2012); Sanches et al. (2013)
Sporophila leucoptera (Vieillot, 1817)	Alo; Acj		8		Luz et al. (2012); Santolin et al. (2012)
Sporophila angolensis (Linnaeus, 1766)	Alo		1		Ogrzewalska et al. (2012)
Poospiza lateralis (Nordmann, 1835)	Aau		3		Arzua et al. (2005)
Volatinia jacarina (Linnaeus, 1766)	Acj		1		Santolin et al. (2012)
Zonotrichia capensis (Statius Muller, 1776)	Aau		1		Arzua et al. (2005)
Cardinalidae Ridgway, 1901					
Habia rubica (Vieillot, 1817)	Alo; Ano	11	5		Labruna et al. (2007); Ogrzewaslka et al. (2009); Ogrzewaslka et al. (2012)
	Asp	12			Labruna et al. (2007); Ogrzewalska et al. (2012)
Cyanoloxia brissonii (Lichtenstein, 1823)	Alo; Aov		3		Arzua et al. (2005); Pacheco et al. (2012)
Parulidae W, F, L, M, P, van Rossem, V T & Z 1947					
Basileuterus culicivorus (Deppe, 1830)	Alo; Ano	16	15		Arzua et al. (2005); Labruna et al. (2007); Ogrzewaslka et al. (2009); Luz et al. (2012); Pacheco et al. (2012)
	Asp	12			Labruna et al. (2007); Ogrzewalska et al. (2012)
Basileuterus hypoleucus Bonaparte, 1830	Alo;		2		Tolesano-Pascoli et al. (2010); Sanches et al. (2013)
Basileuterus flaveolus (Baird, 1865)	Alo; Ano		12		Ogrzewaslka et al. (2009);Tolesano-Pascoli et al. (2010); Luz et al. (2012)
Basileuterus leucoblepharus (Vieillot, 1817)	Alo; Aau	4	5		Arzua et al. (2005); Ogrzewaslka et al. (2012)
	Asp	1			Orgzewaslka et al. (2012)
Basileuterus leucophrys Pelzeln, 1868	Alo; Ano		6		Tolesano-Pascoli et al. (2010)
		15	2		Tolesano-Pascoli et al. (2010)
Geothlypis aequinoctialis (Gmelin, 1789)	Alo; Aca		4		Pacheco et al. (2012)
Icteridae Vigors, 1825					
Cacicus cela (Linnaeus, 1758)	Alo				Aragão (1911)*
Fringillidae Leach, 1820					
Euphonia xanthogaster Sundevall, 1834	Alo		1		Ogrzewalska et al. (2011)
Euphonia violacea (Linnaeus, 1758)	Alo		2		Arzua et al. (2005); Pacheco et al. (2012)
Euphonia pectoralis (Latham, 1801)	Alo	1	2		Labruna et al. (2007); Arzua et al. (2005)
	Asp	2			Labruna et al. (2007)

Table 3. (Continued)

Birds	Ticks				References
NON PASERINES					
Struthioniformes Latham, 1790					
Rheidae Bonaparte, 1849					
Rhea americana (Linnaeus, 1758)	Apv		2	4	Teixeira et al. (2008)
Tinamiformes Huxley, 1872					
Tinamidae Gray, 1840					
Galliformes Linnaeus, 1758					
Cracidae Rafinesque, 1815					
Penelope obscura Temminck, 1815	Alo			2	Arzua et al. (2005)
Penelope superciliaris Temminck, 1815	Abr; Aca		1		Aragão (1911)*; Ogrzewalska et al. (2009);
Pelecaniformes Sharpe, 1891					
Threskiornithidae Poche, 1904					
Theristicus caudatus (Boddaert, 1783)	Acj		4		Labruna et al. (2007)
Accipitriformes Bonaparte, 1831					
Accipitridae Vigors, 1824					
Heterospizias meridionalis	Acj		1		Spolidoro et al. 92012)
Falconiformes Bonaparte, 1831					
Falconidae Leach, 1820					
Caracara plancus (Miller, 1777)		4	5	12	Texeira et al. (2008)
Cariamiformes Furbringer, 1888					
Cariamidae Bonaparte, 1850					
Cariama cristata (Linnaeus, 1766)	Alo; Acj; Apc	40	14		Labruna et al. (2007); Teixeira eta al. (2008)
Charadriiformes Huxley, 1867					
Charadriidae Leach, 1820					
Vanellus chilensis (Molina, 1782)	Acj		10		Santolin et al. (2012)
Columbiformes Latham, 1790					
Columbidae Leach, 1820					
Leptotila verreauxi Bonaparte, 1855	Apv; Aco	10	5		Ogrzewalska et al. (2009); Luz et al. (2012)
Patagioenas cayennensis (Bonnaterre, 1792)	Asp		1		Labruna et al. (2007)
Leptotila rufaxilla	Alo		1		Luz et al. (2012)
	Asp	4			Luz et al. (2012)
Columbina talpacoti (Temminck, 1811)	Acj		1		Santolin et al. (2012)
Geotrygon violacea (Temminck, 1809)	Acj	7			Ogrzewalska et al. (2011)
	Asp	3			Ogrzewalska et al. (2011)

Birds	Ticks				References
Strigiformes Wagler, 1830					
Strigidae Leach, 1820					
Pulsatrix koeniswaldiana (Bertoni & Bertoni, 1901)	Alo		1		Arzua et al. (2005)
Glaucidium brasilianum (Gmelin, 1788)	Asp	1			Ogrzewalska et al. (2009)
Glaucidium minutissimum (Wied, 1830)	Asp	1			Ogrzewalska et al. (2009)
Caprimulgiformes Ridgway, 1881					
Caprimulgidae Vigors, 1825					
Hydropsalis albicollis (Gmelin, 1789)			1	1	Luz et al. (2012)
Apodiformes Peters, 1940					
Trochilidae Vigors, 1825					
Eupetomena macroura (Gmelin, 1788)	Alo		2		Luz et al. (2012)
	Asp	6			Luz et al. (2012)
Chlorostilbon lucidus (Shaw, 1812)	Alo		1		Ogrzewalska et al. (2012)
Thalurania glaucopis (Gmelin, 1788)	Alo	1	1		Ogrzewalska et al. (2011); Ogrzewalska et al. (2012)
	Asp	5			Labruna et al. (2007); Ogrzewalska et al. (2012)
Coraciiformes Forbes, 1844					
Momotidae Gray, 1840					
Baryphthengus ruficapillus (Vieillot, 1818)	Alo; Ano; Acj; Aca; Aco		12		Arzua et al. (2005);Ogrzewalska et al. (2009)
	Asp	1			Ogrzewalska et al. (2009)
Momotus momota	Acj; Aco		2		Ogrzewalska et al. (2009)
Galbuliformes Fürbringer, 1888					
Bucconidae Horsfield, 1821					
Malacoptila striata (Spix, 1824)	Aco; Alo	13	4		Labruna et al. (2007) ; Ogrzewalska et al. (2008); Ogrzewalska et al. (2009); Ogrzewalska et al. (2012)
Malacoptila rufa	Asp	1			Ogrzewalska et al. (2010)
Galbulidae Vigors, 1825					
Galbula cyanicollis Cassin, 1851	Asp	2			Ogrzewalska et al. (2010)
Piciformes Meyer & Wolf, 1810					
Picidae Leach, 1820					
Celeus flavescens (Gmelin, 1788)	Alo		1		Arzua et al. (2005)
Picumnus cirratus Temminck, 1825	Asp	4			Labruna et al. (2007)
Ramphastidae Vigors, 1825					
Ramphastos dicolorus Linnaeus, 1766	Alo		2		Soares et al. (2009)
Ramphastus vitellinus Lichtenstein, 1823	Alo		1		Dantas-Torres et al. (2010)

Table 3. (Continued)

Birds	Ticks				References
Selenidera maculirostris (Lichtenstein, 1823)	Asp	5			Labruna et al. (2007)
Cuculiformes Wagler, 1830					
Cuculidae Leach, 1820					
Crotophaginae Swainson, 1837					
Crotophaga ani Linnaeus, 1758	Alo		1		Pacheco et al. (2012)
Total		**4109**	**1616**	**26**	

Table 4. The species of *Ixodes* associated with wild birds in Brazil. Tick stages (LL= larva, NN= nymph, AA= adult). Ixau= *I. auritulus*, Ixpar= *I. paranaensis*, Ixfus= *I. fuscipes*

Birds	Ticks				References
PASSERIFORMES Linnaeus, 1758	Species	LL	NN	AA	
Thamnophilidae Swainson, 1824					
Thamnophilus ruficapillus Vieillot, 1816	Ixau; Ixpar			4F	Cooley & Kohls (1945); Arzua & Barros-Battetsi (1999); Arzua et al. (2005)
Thamnophilus caerulescens Vieillot, 1816	Ixau; Ixfus		1	5F	Cooley & Kohls (1945)
Drymophila malura (Temminck, 1825)	Ixfus		1		Arzua et al. (2005)
Conopophagidae E, O, Irested, Alvarenga & Fjeldsa, 2010					
Conopophaga lineata (Wied, 1831)	Ixau	2	2		Arzua & Barros-Battesti (1999); Arzua et al. (2005)
Dendrocolaptidae Gray, 1840					
Xiphorhynchus fuscus (Vieillot, 1818)	Ixfus		1		Arzua et al. (2005)
Campylorhamphus falcularius (Vieillot, 1822)	Ixfus		1		Arzua et al. (2005)
Furnariidae Gray, 1840					
Syndactyla rufosuperciliata (Lafresnaye, 1832)	Ixau;Ixfus		11		Arzua & Barros-Battesti (1999); Arzua et al. (2005)
Synallaxis ruficapilla Vieillot, 1819	Ixau	2	2		Arzua et al. (2003); Arzua et al. (2005)
Clibanornis dendrocolaptoides (Pelzeln, 1859)	Ixau	1	2	1F	Arzua et al. (2005)
Turdidae Rafinesque, 1815					

Birds	**Ticks**				**References**
Turdus rufiventris Vieillot, 1818	Ixau	151	112	93F	Arzua et al. (1994); Arzua & Barros-Battesti (1999); Arzua et al. (2003); Arzua et al. (2005); Amaral et al. (2013)
Turdus amaurochalinus Cabanis, 1850	Ixau		6		Arzua & Barros-Battesti (1999); Arzua et al. (2005); Amaral et al. (2013)
Turdus albicollis Vieillot, 1818	Ixau		2	4F	Arzua et al. (1994); Arzua & Barros-Battesti (1999); Arzua et al. (2005)
Thraupidae Cabanis, 1847					
Lanio melanops (Vieillot, 1818)	Ixfus		2		Arzua et al. (2005)
Tachyphonus coronatus (Vieillot, 1822)	Ixfus		2		
Emberizidae Vigors, 1825					
Poospiza lateralis (Nordmann, 1835)	Ixau		2		Arzua et al. (2005)
Haplospiza unicolor Cabanis, 1851	Ixau	1			Arzua et al. (2005)
Tyrannidae Vigors, 1825					
Knipolegus nigerrimus (Vieillot, 1818)	Ixau			1F	Cooley & Kohls 1945
Parulidae W, F, L, M, P, R, Van Tyne & Zimmer 1947					
Basileuterus leucoblepharus (Vieillot, 1817)	Ixau	1			Arzua et al. (2005)
Basileuterus culicivorus (Deppe, 1830)	Ixfus	1			Arzua et al. (2005)
NÃO PASERIFORMES					
Columbiformes Latham, 1790					
Columbidae Leach, 1820					
Columbina talpacoti (Temminck, 1811)	Ixau			2F	Arzua & Barros-Battesti (1999); Arzua et al. (2005)
Apodiformes Peters, 1940					
Apodidae Olphe-Galliard, 1887					
Streptoprocne biscutata (Sclater, 1866)	Ixpar	2	17	1M 9F	Arzua et al. (2005); Barros-Battesti et al. (2003)
Streptoprocne zonaris	Ixpar		4	1F	Arzua & Barros-Battesti (1999); Arzua et al. (2005); Barros-Battesti et al. (2003)
Total		161	168	1M 120F	

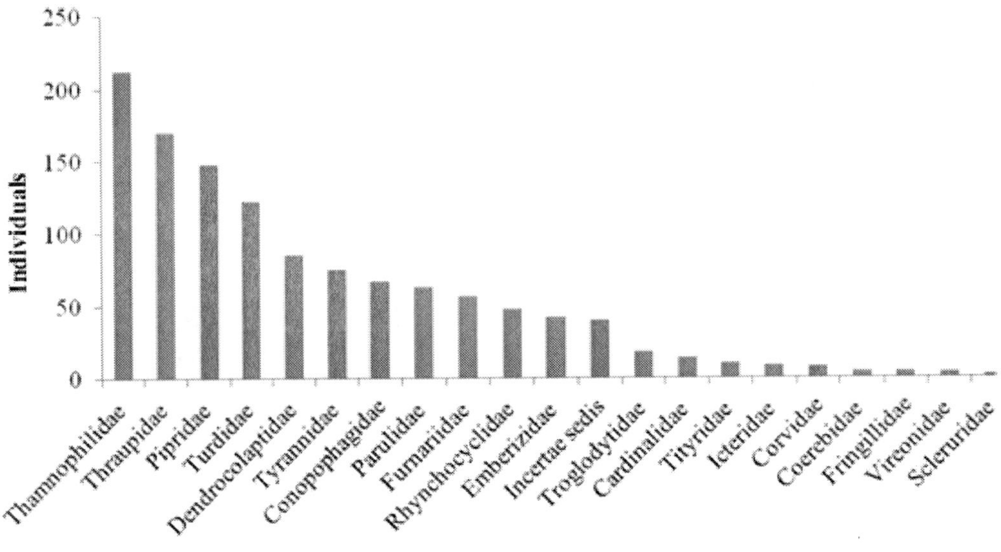

Figure 2. Number of individuals of passerine birds parasitized by ticks in Brazil.

The species *A. longirostre*, popularly known as the bird tick or bird earring (Aragão, 1936) is the most common species. It is currently recorded from 71.6% (116/163) passerine species comprising 18 families (Table 3). Among the bird species, *Tachyphonus coronatus* (Vieillot) (Ruby-crowned Tanager) (Thraupidae) and *Turdus rufiventris* Vieillot (Rufous-bellied Thrush) (Turdidae) are the most parasitized. Both species live in the middle and high strata of forest trees, so these birds can coexist with arboreal mammals, a group with individuals commonly parasitized by *A. longirostre* as the rodents Erethizontidae (Brum et al., 2003; Labruna et al., 2002; 2004; Silveira et al., 2008). Labruna et al. (2007) suggested that *A. longirostre* is an arboreal tick since the free-living stages search porcupines (*Coendou* spp.), primary hosts for the adult stage and arboreal birds, hosts for immatures, in the canopy. The migratory birds *Xolmis cinereus* (Vieillot), *Tyrannus savanna* Vieillot and *Elaenia flavogaster* (Thunberg, 1822) are parasitized by *A. longirostre* (Figure 3).

The species *A. nodosum* is the second species most commonly associated with passerines especially *Turdus amaurochalinus* Cabanis (Creamy-bellied Thrush) (Turdidae) and *Conopophaga lineata* Wied (Rufous Gnateater) (Conopophagidae). It is currently recorded in 30 bird species, distributed in 13 families. *T. amaurochalinus* is a migratory species widely distributed in Brazil (Figure 3). Adults of *A. nodosum* are common parasites of Xenartha (Barros-Batestti, 2006).

The species *A. aureolatum*, the third most common tick, is currently recorded from 13.6 % (22/163) species of passerines, including nine families with two species most parasitized *T. rufiventris* and *Troglodytes musculus* Naumann (Southern House Wren) (Troglodytidae). Hosts for adults are deer, opossums, wild carnivores, domestic dog and man mainly living in rural forested areas (Barros and Baggio, 1992; Evans et al., 2000; Barros-Batestti, 2006). The two most parasitized species of birds occupy different habitats, *T. rufiventris* inhabits high and middle strata of forest trees whereas *T. musculus* inhabits low strata (Sick 1997; Ferreira et al., 2010; Luz et al., 2012), but both species are resistant to anthropogenic impact on the environment. The second species is very common in agricultural areas.

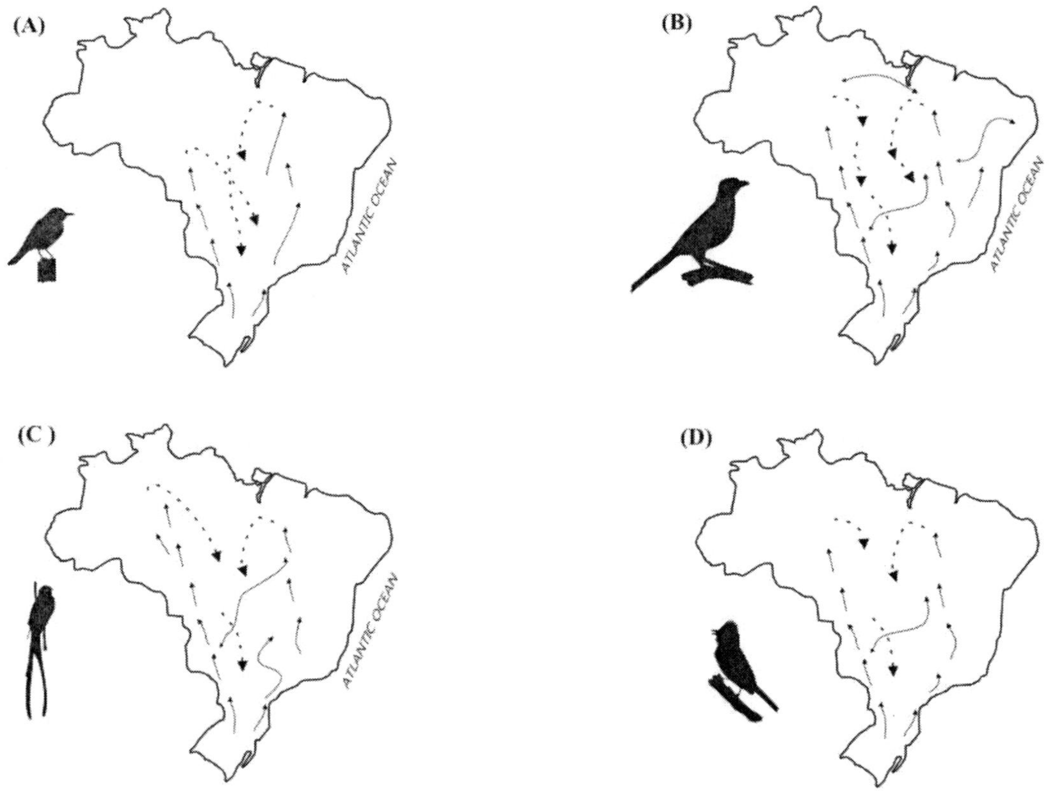

Figure 3. A-D. Migrations routes of four species of parasitized birds edemic to Brazil. A= Xolmis cinereus winter, B= Turdus amaurochalinus, C= Tyrannus savana, D= Elaenia flavogaster.

The genus *Ixodes* bears three species: *Ixodes auritulus* Neumann, *Ixodes paranaensis* Barros-Battesti, Arzua, Pichorim & Keirans and *Ixodes fuscipes* Koch. The tick *I. auritulus* is the most common in passerines, being recorded in eight families and 11 species of birds. The bird *T. rufiventris* is the most parasitized species with 151 larvae, 112 nymphs and 93 females recorded (Table 4).

Some species of the genus *Ixodes*, primarily adults, parasitizes preferably vertebrates as pacas, opossums and agoutis, while the immature stages can be found in small mammals, especially rodents (Guimarães et al., 2001; Díaz et al., 2009; Luz et al., 2013). In Brazil, *I. auritulus* can be found mainly in the southeastern and southern regions (Arzua and Barros-Batestti, 1999).

The remaining species *A. coelebs*, *A. tigrinum, A. pseudoconcolor, A. naponense*, *I. paranaensis* and *I.fuscipes are recorded sporadically from* passerine birds (Aragão, 1936; Evans, 2000; Arzua et al., 2005; Barros-Battesti et al., 2003; Labruna et al., 2007; Sazbó et al., 2008; Luz et al., 2012; Orgzewalska et al., 2009; 2010; 2012).

Two species of the genus *Haemaphysalis* are recorded from passerines in Brazil: *Haemaphysalis leporispalustris* Packard, one larva in *Lanio melanops* (Vieillot) (Black-goggled Tanager) and *Haemaphysalis juxtakochi* Cooley, one larva in *Hylophylax naevius* (Gmelin) (Spot-backed Antbird), two nymphs in *Willisornis poecilinotus* (Cabanis) (Scale-backed Antbird) and three adults in *Pyrrhocoma ruficeps* (Strickland, 1844) (Chestnut-headed Tanager). *R.sanguineus* is recorded in *C. flaveola* (1 male) and *Gnorimopsar chopi* (Vieillot)

(Chopi Blackbird) (7 nymphs) but this may be an accidental record due to the closeness of these birds and dogs in rural areas (Luz et al., 2012).

Relationship between Ticks and Non-Passerine Birds in Brazil

Table 3 and 4 shows details of parasitism by species of *Amblyomma* and *Ixodes* in non-passerine birds. The non-passerines are also quite diverse morphologically and ecologically in Brazil. Briefly, they can be classified in four groups regarding their habitats: i) inhabiting rivers and lakes such as *Egretta thula* (Molina), (Snowy Egret), *Phalacrocorax brasilianus* (Molina) (Neotropic Cormorant) and *Butorides striata* (Linnaeus) (Striated Heron), ii) forestry such as *Galbalcyrhynchus leucotis* Des Murs (White-eared Jacamar), *Monasa atra* (Boddaert) (Black Nunbird) and *Ramphastos vitellinus* (Lichtenstein, 1823) (Channel-billed Toucan), iv) coastal environments such as *Larus dominicanus* Lichtenstein (Kelp Gull) and *Sula leucogaster* (Boddaert) (Brown Booby) and iv) fresh or brackish waters such as *Calidris melanotos* (Vieillot) (Pectoral Sandpiper) and *Tringa solitaria* Wilson (Solitary Sandpiper) (Sick ,1985; 1997; Sigrist, 2007).

These birds are included into 30 orders, 70 families and 809 species. The Psittacidae (85 species), Trochilidae (83 species) and Picidae (51 species) are the most common in Brazil (CBRO 2011). Overall, 67 species are endemic and distributed across all biomes (Sick 1977; Marini and Garcia 2005; CBRO 2011). To date, parasitism by ticks in non- passerine birds is recorded in approximately 53.3% (16/30) orders, 25.7% (18/70) families and 4.2% (34/809) species. The Columbiformes and Apodiformes are the groups with higher number of bird species parasitized. In Columbiformes, there are five species parasitized in Columbidae whereas in Apodiformes there are three species parasitized in Trochilidae and two bird species parasitized in Apodidae.

Of the 83 individual birds parasitized, 21 belong to Apodidae, 11 to Momotidae and 10 to Charadriidae (Figure 4). In the Apodidae, *Streptoprocne biscutata* (Sclater) (Biscutate Swift) is the most parasitized bird.

Overall, approximately 237 ticks distributed as larvae (110, 46%), nymphs (95, 40%) and adults (32, 13%) (Table 2) and included in seven tick species and in two genera *Amblyomma* and *Ixodes*. The genus *Amblyomma*, as in passerines, is the most diverse, including six species: *A. longirostre*, *A. cajennense*, *A. pacae*, *A. parkeri* and *A. parvum* (Table 3). The species *A. pacae*, *A. cajennense* and *A. longirostre* are the most abundant with 50 ticks (40 larvae and 10 nymphs), 45 ticks (11 larvae, 22 nymphs and 12 females) and 33 ticks (14 larvae, 16 nymphs and 3 females), respectively.

The tick *A. longirostre* is identified in 36 % (13/34) bird species. Three species of Apodiformes, *Eupetomena macroura* (Gmelin) (Swallow-tailed Hummingbird), *Chlorostilbon lucidus* (Shaw) (Glittering-bellied Emerald) and *Thalurania glaucopis* (Gmelin) (Violet-capped Woodnymph) are the most parasitized. However, the sole parasitized species of Galbuliformes (*Malacoptila striata* (Spix) - Crescent-chested Puffbird) showed the greatest abundance (13 larvae and 3 nymphs) by this tick species.

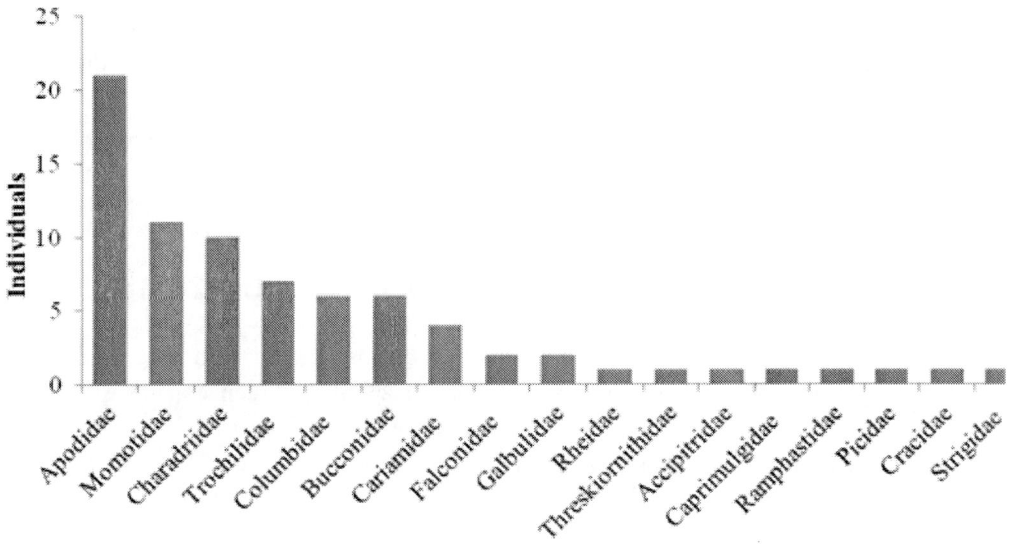

Figure 4. Number of individualsof non passerine birds parasitized by ticks in Brazil.

The association of *A. longirostre* with hummingbirds is rather interesting and needs to be better studied. Hummingbirds live primarily in forests but some species are adapted to urban centers. Studies with ringed hummingbirds showed bird flights from two to 30 km in one day, reaching altitude of approximately 455m to 1400m (Sick, 1997). This behavior may imply these birds as possible carrier of *A. longirostre* for long distances and high altitudes.

The relationship between *M. striata* (Buconidae) and *A. longirostre* is also worthy of attention. Although nesting in bounds, *M. striata* belongs to a group of so-called arboreal birds (Sick, 1997). This behavior may also favor parasitism by *A. longirostre*, supposedly an arboreal tick (Labruna et al., 2007). This bird is distributed mainly in the southeastern and southern states but also occurs and in some northeastern states. However, its association with *A. longirostre*, so far, is restricted to the southeast.

The tick *A. cajennense* is currently known from 23% (8/34) non passerine species included in six families (Table 1). Most are grounding foragers (Sick, 1997) which may explain the frequent occurrence of *A. cajennense* (Labruna et al., 2007; Teixeira et al., 2008; Luz et al., 2012; Santolin et al., 2012) in these birds but additional studies are needed to confirm this association. *Caracara plancus* (Miller) (Southern Caracara) is the most parasitized bird (4 larvae, 5 nymphs and 12 females).

To date *A. pacae* is recorded in only one bird species, *C. cristata* (40 larvae and 10 nymphs) (Teixeira et al., 2008). *C. cristata* is widely distributed in Brazil, absent only in the Amazon forest. It is a typical bird of the Cerrado biome which forages on the ground and migrates for long distances.

Two species of *Ixodes* are currently associated with non-passerines: *I. auritulus* (2 females) in *Columbina talpacoti* (Ruddy Ground-Dove) and *I. paranaensis* (2 larvae, 13 nymphs, 9 adults) in *S. biscutata* (Biscutate Swift) and four nymphs and one adult in *Streptoprocne zonaris* (Shaw) (White-collared Swift). Additional research on the association between ticks and the two species of *Streptoprocne* is of great importance since they can migrate through the Brazilian biomes (Sick, 1997).

According Arzua et al., (2003), *I. auritulus* is a common parasite of Turdidae, particularly *T. rufiventris*. The finding of this tick in *C. talpacoti* can be regarded as accidental until additional studies clarify the issue.

The species *A. parkeri*, *A. parvum*, *A.ovale* and *A. brasiliensis* are recorded sporadically in association with non-passerine birds

Seasonality and Spread of Pathogens

Data on seasonality and spread of pathogens by ticks associated with birds in Brazil are scarce and fragmented. Regarding seasonality few surveys have been carried out, so that, currently it is impossible to establish seasonal patterns. Surveys carried out in the biomes Atlantic Forest (Marini et al., 1996; Rojas et al., 1999; Storni et al., 2005) and Cerrado (Luz et al., 2012) have shown more immatures of *Amblyomma* sp.in the winter season (dry/cold). Surveys conducted in the city of Curitiba, south Brazil, have shown occurrence of larvae of *I. auritulus* in fall and winter, nymphs in fall and spring, and females in spring (Arzua et al., 1999; 2003).

Ticks are of great concern of Veterinary and Public Health welfare because they cause direct damage during feeding and act as vectors of several infectious agents to humans, domestic animals and pets (Barros- Battesti et al., 2006) throughout the world.

Of the 21 species associated with birds in Brazil, five are of veterinary and public health concern: *A. cajennense*, *A. aureolatum*, *A. ovale, A. parvum* and *R. sanguineus*. For instances, the four species of *Amblyomma* are involved in the transmission of the Brazilian Spotted Fever complex whereas the last species is the vector of Canine Ehrlichiosis.

A. cajennense and *A. aureolatum* are the most important vectors of the Brazilian strains of *Rickettsia rickettsii* to humans and animals (Pinter and Labruna, 2006; Labruna et al., 2011).

To date only bacteria from the genus *Rickettsia* have been isolated from ticks associated with wild birds: from *A. longirostre, Rickettsia amblyommi* and *Rickettsia parkeri* (Ogrzewalska et al., 2008; 2009; 2011; 2012; Pacheco et al., 2012); from *A. nodosum, R. parkeri* (Ogrzewalska et al., 2009; Pacheco et al., 2012); from *A. geayi, R. amblyommi* (Ogrzewalska et al., 2010); from *A. parkeri, R. parkeri* (Ogrzewalska et al., 2012), from *A. calcaratum, R. parkeri* (Ogrzewalska et al., 2013) and from *Amblyomma sp., R. parkeri* (Ogrzewalska et al., 2012). Although knowledge of the association pathogens-ticks-wild birds is still at an embryonic stage in Brazil, the few studies mentioned herein are sufficient to highlight the importance of further research to deeply investigate the real role played by wild birds in the spread of diseases in Brazil, especially the migratory birds.

Conclusion and Future Research

Over the past 10 years there has been an increase in studies on parasitism by ticks on birds in Brazil. However, there are still numerous gaps that need to be filled both qualitatively and quantitatively.

Regarding the capture methods used to date, it is important to draw attention to the following items. i) mist nets are effective for capturing understory birds, however this method of capture overlooks large number of species that inhabit the upper strata of the forest. The exclusive use of this method in most of tick surveys may explain the low captures of non-passerine birds, which usually use the upper strata of the forest. ii) the size of the mesh is also important and many studies used mesh sizes of 30 mm and 36 mm. For instance, the mesh of 36mm is not suitable to capture larger birds such as Columbidae.

We also draw attention to the lack of bird banding in the past surveys as determined by the Committee of Brazilian Ornithological Records (CBRO in Portuguese) and the National Centre for Research and Conservation of Wild Birds (CEMAVE in Portuguese). The banding is an individual marking of the bird, its identity. With the marking of individual birds, we can build a database of information required for future decisions.

Concomitant investigations on ticks parasitizing vertebrates, mainly small arboreal mammals which inhabit the tree canopies are needed as well. There is not such information in Brazil.

Also of concern are the migratory birds that carry ticks to new areas, especially those which migrate from North and Central America. Such studies are important to clarify the etiology of emerging and re-emerging diseases transmitted by ticks which occur in Brazil. With regard to pathogens itself, it is necessary to evaluate the role of birds in the maintenance and amplification of pathogens, their ability to infect ticks and further the transmission to other groups of vertebrates, especially mammals, which are mainly hosts of adult stage of tick species that parasitize birds and are part of epidemiological cycle of pathogens transmitted by ticks.

Finally, the few existing studies on ticks seasonality mentioned in this chapter are based upon poorly gained data. Therefore, we need additional research efforts to clarify the pattern of seasonal variation of tick parasitism.

References

Amaral, H. L. C., Bergmann, F. B., Silveira, P. R. S., Krüger, R. F., & Graciolli, G. (2013). Community of arthropod ectoparasites of two species of *Turdus* Linnaeus, 1758 (Passeriformes: Turdidae) in southern Rio Grande do Sul, Brazil. *Parasitology Research*, *112*, 621–628.

Aragão H. (1936). Ixodidas brasileiros e de alguns paizes limitrophes. *Memórias* do *Instituto Oswaldo Cruz*, *31*, 759–843.

Aragão, H. B., Fonseca, F. (1961). Notas de ixodologia. *Mem. Inst. Oswaldo Cruz,* 59, 119-129.

Arzua, M., & Barros-Battesti, D. M. (1999). Parasitism of *Ixodes* (*Multidentatus*) *auritulus* Neumann (Acari: Ixodidae) on birds from the City of Curitiba, State of Paraná, Southern Brazil. *Memórias* do *Instituto Oswaldo Cruz*, *94*, 597–603

Arzua, M., Silva, M. A. N., Famadas, K. M., Beati, L., & Barros-Battesti, D. M. (2003). .*Amblyomma aureolatum* and *Ixodes auritulus* (Acari: Ixodidae) on birds in southern Brazil, with notes on their ecology. *Experimental and Applied Acarology*, *31*, 283-296.

Arzua, M., Onofrio, V. C., & Barros-Battesti, D. M. (2005) Catalogue of the tick collection (Acari: Ixodidae) of the Museu de História Natural Capão da Imbuia, Curitiba, Paraná, Brazil. *Revista Brasileira de Zoologia*, *22*, 623-632.

Barros, D. M., & D. Baggio. (1992). Ectoparasites Ixodida Leach 1817 on wild mammals in the State of Paraná, Brazil. *Memórias do Instituto Oswaldo Cruz*, *87*, 291-296.

Barros-Battesti, D. M., & Knysak, I. (1999). Catalogue of the Brazilian *Ixodes* (Acari: Ixodidae) material in the mite collection of the Instituto Butantan, São Paulo, Brazil. *Papéis Avulsos Zoologia 41*: in press.

Barros-Battesti, D. M., Arzua, M., Pichorim, M., & Keirans, J. E. (2003). *Ixodes* (*Multidentatus*) *paranaensis* n. sp. (Acari: Ixodidae) a Parasite of *Streptoprocne biscutata* (Sclater 1865) (Apodiformes: Apodidae) Birds in Brazil. *Memórias do Instituto Oswaldo Cruz*, *98*, 93-102, 2003.

Barros-Battesti, D. M., Arzua, M., & Bechara, G. H. (2006) Carrapatos de importância médico-veterinária da Região Neotropical: Um guia ilustrado para identificação de espécies. Vox/ICTTD-3/Butantan, São Paulo.

Brum, J. G. W., Valente, A. L. S., & Albano, A. P. (2003). Ixodidae de mamíferos silvestres atendidos no Núcleo de Reabilitação da Fauna Silvestre, UFPEL. *Arquivos Instituto Biológico, 70*, 211-212.

Comitê Brasileiro de Registros Ornitológicos – CBRO (2011). *Epi Info* [online]. Available from: http://www.cbro. org.br/CBRO/index.htm. Cited 15 Jun 2013.

Daniels, T. J, Battaly, G. R, Liveris D, Facol, R. C., & Schwartz., I. (2002). Avian reservoirs of the agent of human granulocytic ehrlichiosis? *Emerging Infectious Diseases, 8*, 1524–1525.

Dantas-Torres, F., Onofrio, V.C., Barros-Battesti. (2009). The ticks (Acari: Ixodida: Argasidae,Ixodidae) of brazil. *Systematic & Applied Acarology, 14*, 30-46.

Díaz, M. M., Nava, S., Guglielmone, A. A. (2009). The parasitism of *Ixodes luciae* (Acari: Ixodidae) on marsupials and rodents in Peruvian Amazon. *Acta Amazonica*, *39*, 997-1002.

Elfving, K., Oslen, B., Bergström, S., Waldenström, J., Lundkvist, A., Sjöstedt., Mejlon, H. Nilsson, K. (2010). Dissemination of Spotted Fever Rickettsia agents in Europe by migrating birds. *PLoSOne*, *5*,8572.

Evans, D. E., Martins, J. R., Guglielmone, A. A. (2000). A review of the ticks (Acari, Ixodida) of Brazil, their hosts and geographical distribution – 1. The state of Rio Grande do Sul, southern Brazil. *Memórias do Instituto Oswaldo Cruz*, *95*, 453–470.

Faccini, J. L. H., Barros-Battesti, D. (2006). Comentários e chaves para as espécies do gênero *Amblyomma.* In Barros-Battesti DM, Arzua M, Bechara GH (eds) Carrapatos de importância médico-veterinária da Região Neotropical: um guia ilustrado para identificação de espécies. Vox/ICTTD-3/Butantan, São Paulo, 53-113.

Ferreira, I., Ventura, P. E. C., Luz, H . R. (2010). Aves no campus da UFRRJ. EDUR.

Guglielmone AA, Estrada-Peña A, Keirans JE, Robbins RG. 2003. Ticks (Acari: Ixodida) of the Neotropical Zoogeographic Region. Atalanta, Hauten: International Consortium on Ticks and Tick-Borne Diseases (ICTTD-2), 173 p.

Guglielmone, A. A., Robins, R. G., Apanaskevich, D. A., Petney, T. N., Estrada-Peña, A., Horak, I. G., Shao, R., Barker, S. C. (2010). The Argasidae, Ixodidae and Nuttalliellidae (Acari: Ixodida) of the world: a list of valid species names. *Zootaxa*, 2528, 1-28.

Guimaraes, J. H., Tucci, E. D., Barros-Battesti, D. M. (2001). Ectoparasitos de Importância Veterinária. Pleiade-FAPESP, São Paulo.

Graham, R. I., Mainwaring, M. C., Du Feu, R. (2010). Detection of spotted fever group *Rickettsia* spp. from bird ticks in the U.K. *Medical and Veterinary Entomology, 24*, 340-343.

Hasle, G., Bjune, G., Edvardsen, E., Jakobsen, C., Linnehol, B., Røer, J., Mehl, R., Røed, K., Pedersen, J., Leinas, H. (2009). Transport of ticks by migratory passerine birds to Norway. *Journal Parasitology*, *95*, 1342-1351.

Hoogstraal, H. (1961). Migrating birds and their ectoparasites in relation to disease. *East African Medical Journal*, *38*, 221-238.

Hubalek, Z. (2004) An annotated checklist of pathogenic microorganisms associated with migratory birds. *Journal of Wildlife Diseases*, *40*, 639-659.

Ioannou, I., Chochlakis, D., Kasinis, N., Anayiotos, P., Lyssandrou, A., Papadopoulos, B., Tselentis, Y., Psaroulaki, A. (2009). Carriage of *Rickettsia* spp., *Coxiella burnetii* and *Anaplasma* spp. By endemic and migratory wild birds and their ectoparasites in Cyprus, *European Society of Clinical Microbiology and Infectious Diseases*, *4*, 1-3.

Jordan, B. E., Onks, K. R., Hamilton, S. W., Hayslette, S. E., Wright, S. M. (2009). Detection of *Borrelia burgdorferi* and *Borrelia lonestari* in Birds in Tennessee. *Journal of Medical Entomology*, *46*, 131-138.

Labruna, M.B., Paula, C.D., Lima, T.F. Sana, D. A. (2002) Ticks (Acari: Ixodidae) on wild animals from the Porto-Primavera hydroelectric power station area, Brazil. *Memórias do Insituto Oswaldo Cruz, 97*, 1133-1136.

Labruna, M.B., Mcbride, J.W., Bouyer, D.H., Camargo, L. M., Walker, D. H. (2004). Molecular evidence for a spotted fever group *Rickettsia* species in the tick *Amblyomma longirostre* in Brazil. *Journal of Medical Entomology, 41*, 533-537.

Labruna, M. B., Wlippo, L. F. S., Demetrio, C., Menezes, A. C., Pinter, A., Guglielmone, A. A., Silveira, L. F. (2007). Ticks collected on birds in the state of São Paulo, Brazil. *Experimental and Applied Acarology*, *43*, 147-160.

Labruna, B., Mattar, S., Nava, S., Bermudez, S., Venzal, J. M., Dolz, G., Abarca, K. M. D., Romero, L., Sousa, R., Oteo, J.M. D., Zavala-Castro, J. (2011). Rickettsioses in Latin America, Caribbean, Spain and Portugal, *Revista Medicina Veterinaria y Zootecnia Córdoba, 16*, 2435-2457.

Luz, H. R., Faccini, J. L. H., Landulfo, G. A., Sampaio, J. S., Neto, S. F. C., Famadas, K. M., Onofrio, V. C., Barros-Battesti, D. M. (2013). New host records of *Ixodes luciae* (Acari: Ixodidae) in the State of Pará, Brazil. Revista Brasileira de Parasitologia Veterinária, *22*, 152-154.

Luz, H. R., Faccini, J. L. H., Landulfo, G. A., Berto, B. P., Ferreira, I. (2012). Bird ticks in an area of the Cerrado of Minas Gerais State, southeast Brazil. *Experimental and Applied Acarology*, *58,* 89-99.

Martins, T. F., Onofrio, V. C., Barros-Battesti, D. M., Labruna, M. B. (2010). Nymphs of the genus *Amblyomma* (Acari: Ixodidae) of Brazil: Descriptions, redescriptions, and identification key. *Ticks Tick Borne Disease* 2010, *1*, 75-99.

Marini, M. A., Reinert, B. L., Bornschein, M. R., Pinto, J. C., Pichorim, M. A. (1996). Ecological correlates of ectoparasitism on Atlantic Forest birds, Brazil. *Ararajuba, 14*, 93–103.

Marini, M. A., Garcia, F. I. (2005) Conservação de aves no Brasil, *Megadiversidade,1,*1. , 95-102.

Ogden, N.H., Lindsay, L.R., Hanincová, K., Barker, I.K., Bigras-Poulin, M., Charron, D.F., Heagy, A., Francis, C.M., O'Callaghan, C.J., Schwartz, I., Thompson, R.A., (2008). Role of Migratory Birds in Introduction and Range Expansion of Ixodes scapularis Ticks and of *Borrelia burgdorferi* and *Anaplasma phagocytophilum* in Canada. *Applied and Environmental Microbiology*, *74*, 1780-1790.

Ogrzewalska, M., Pacheco, R., Uezu, A., Ferreira, F., Labruna, M. B. (2008).Ticks (Acari: Ixodidae) infesting wild birds in an Atlantic Forest area in the state of São Paulo, Brazil, with isolation of *Rickettsia* from the tick *Amblyomma longirostre*. *Journal of Medical Entomology*, *45*, 770-774.

Ogrzewalska, M., Pacheco, R. C., Uezu, A., Richtzenhain, L. J., Ferreira, F., Labruna, M. B. (2009). Ticks (Acari: Ixodidae) infesting birds in an Atlantic rain forest region of Brazil. *Journal of medical Entomology*, *46*, 1225-1229.

Ogrzewalska, M., Uezu, A., Labruna, M. B. (2010). Ticks (Acari: Ixodidae) infesting wild birds in the eastern Amazon, northern Brazil, with notes on rickettsial infection in ticks. Parasitology Research, *106*, 809–816.

Ogrzewalska, M., Uezu, A., Labruna, M. B. (2011). Ticks (Acari: Ixodidae) infesting wild birds in the Atlantic Forest in northeastern Brazil, with notes on rickettsial infection in ticks. *Parasitology Research*, *108*, 665-670.

Ogrzewalska, M., Saraiva, D. G., Moraes-Filho, J., Martins, T. F., Costa, F. B., Pinter, A., Labruna, M. B. (2012). Epidemiology of Brazilian spotted fever in the Atlantic Forest, state of São Paulo, Brazil, *Parasitology*, 1 -18.

Ogrzewalska, M., Martins, T., Capekb, M., Literakc, I., Labruna, M. B. (2013). A Rickettsia parkeri-like agent infecting *Amblyomma calcaratum* nymphs from wild birds in Mato Grosso do Sul, Brazil. *Ticks and Tick-borne Diseases*, *4*, 145– 147.

Onofrio, V. C., Barros-Battesti, D. M., Labruna, M. B., Faccini, J. L. (2009). Diagnoses of and illustrated key to the species of Ixodes Latreille, 1795 (Acari: Ixodidae) from Brazil. *Systematic Parasitology*, *72*, 143-157.

Onofrio, V. C., Venzal, J. M., Pinter, A., Szabó, M. P. J. Família Ixodidae: características gerais, comentários e chaves para gêneros. In: Barros- Battesti DM, Arzua M, Bechara GH. *Carrapatos de importância medico-veterinária da região neotropical*: um guia ilustrado para identificação de espécies. São Paulo: Vox/ICTTD-3, Butantan, 2006. p. 29-39.

Pacheco, R. C., Arzua, M., Nieri-Bastos, F. A., Moraes-Filho, Jonas, A.. Leonardo, M., Richtzenhain, J., Barros-Battesti, D. M., Labruna, M. B. (2012). Rickettsial Infection in Ticks (Acari: Ixodidae) Collected on Birds in Southern Brazil. *Medical Entomology*, *49*, 716.

Pascoal., J. O., Amorim, M. P., Martins, M. M., Melo, C., Júnior, E. L. S., Ogrzewalska, M., Labruna, M. B., Szabó, M. P. J. (2012). Ticks on birds in a savanna (Cerrado) reserve on the outskirts of Uberlândia, Minas Gerais, Brazil, Revista *Brasileira de Parasitologia Veterinária*, *22*, 46-52.

Pinter, A., Labruna, M. B. (2006). Isolation of *Rickettsia rickettsii* and *Rickettsia bellii* in cell culture from the tick *Amblyomma aureolatum* in Brazil. *Annals of the New York Academy of Sciences*, *1078*, 523-530.

Rojas, R., Marini, M. A., Zanatta, C. A. T (1999) Wild birds as hosts of *Amblyomma cajennense* (Fabricius, 1787) (Acari: Ixodidae). *Memórias do Instituto Oswaldo Cruz, 94,* 315–322.

Sanches, G. S., Martins, T. F., I Lopes, T., Costa, L. F. S., Nunes, P. H., Camargo-Mathias, M. I., Labruna, M. B. (2013). Ticks infesting birds in Atlantic Forest fragments in Rio Claro, State of Sao Paulo, Brazil, *Brasileira de Parasitologia Veterinária, 22,* 6-12.

Santolin, Í. D. A. C., Luz, H. R., Alchorne, N. M., Pinheiro, M. C., Melinski, R. D., Faccini, J. L. H., Ferreira, I., Famadas, K. M. (2012) Ticks on birds caught on the campus of the Federal Rural University of Rio de Janeiro, *Brasileira de Parasitologia Veterinária,* 21, 213-218.

Santos-Silva, M. M., Sousa, R., Santos, A. S., Melo, P., Encarnação, V., Bacellar, F. (2006). Ticks parasiting wild birds in Portugal: detection of *Rickettsia aeschlimanii, R. helvética* and *R. massiliae*. *Experimental and Applied Acarology*, *39*, 331-33.

Sick, H. (1997). *Ornitologia Brasileira*. Editora Nova Fronteira. 504p.

Sigrist, T. (2007). Guia de campo, aves do Brasil Oriental. Avisbrasilis, São Paulo, 448 p.

Silveira, J. A. G., Oliveira, P. A., Curi, N. H. A., Barata, R. S. L., Chiarello, A. G., Ribeiro, M. F. B. (2008). Ocorrência de *Amblyomma longirostre* (Koch, 1844) em *Chaetomys subspinosus* (Olfers, 1818) oriundos da Mata Atlântica. *Arquivo Brasileiro de Medicina Veterinária e Zootecnia*, *60,* 3.

Scott, J. D., Fernando, K., Banerjee, S. N., Durden, L. A., Byrne, S. K., Banerjee, M., Mann, R. B., Morshed, M. G. (2001). Birds disperse ixodid (Acari: Ixodidae) and Borrelia burgdorferi-infected ticks in Canada. *Journal of Medical Entomology*, *38*, 493-500.

Spolidorio, M. G., M. B. Labruna, E. Mantovani, P. E. Brandao, L. J. Richtzenhain, and N. H. Yoshinari. (2010). Novel spotted fever group rickettsiosis, Brazil. *Emerging Infectious Diseases*, 16, 521-523.

Storni, A., Alves, M. A.S., Valim, M. P. (2005) Ácaros de penas e carrapatos (Acari) associados a *Turdus albicollis* Vieillot (Aves, Muscicapidae) em uma área de Mata Atlântica da Ilha Grande, Rio de Janeiro, Brasil. Revista Brasileira de Zoologia, 22, 419–423.

Szabó, M. P. J., Tolesano – Pascoli, V. G., Júnior, O. M., Franchin, A. G., Torga, K. (2008). Brown dog tick *Rhipicephalus sanguineus* parasitizing the bird *Coereba flaveola* in the Brazilian Cerrado. *Ciência Rural*, 38, 543-545.

Teixeira, R. H. F., Ferreira, I., Amorim, M., Gazeta, G. S., Serra-Freire, N M. (2008). Carrapatos em aves selvagens no Zoológico de Sorocaba – São Paulo, Brasil. *Arquivo Brasileiro de Medicina Veterinária e Zootecnia,* 60, 1277-1280.

Tolesano-Pascoli, G. V., Torga, K., Gabriel, F. A., Ogrzewalska, M., Gerardi, M., Olegário, M. M. M., Labruna, M. B., Szabó, M. P. J., Júnior, O. M. (2010). Ticks on birds in a forest fragment of Brazilian cerrado (savanna) in the municipality of Uberlândia, State of Minas Gerais, Brazil. *Revista Brasileira de Parasitologia Veterinária,* 19, 244-248.

In: Birds
Editors: Lucas Ruiz and Franco Iglesias

ISBN: 978-1-62948-104-3

Chapter 7

An Overview of Recent Parasitic Diseases due to Helminths and Arthropods Recorded from Wild Birds, with Special Reference to Conservation Medical Cases from the Wild Animal Medical Center of Rakuno Gakuen University in Japan

Kii Ushiyama[1], Tomoo Yoshino[2], Takuro Hirayama[1], Yuichi Osa[3] and Mitsuhiko Asakawa[1,4]*
[1]Department of Pathobiology, School of Veterinary Medicine, Rakuno Gakuen University, Ebetsu, Hokkaido, Japan
[2]Kyushu Regional Environment Office, Ministry of the Environment. Kumamoto, Japan (Present: Kushiro Zoo, Kushiro, Hokkaido, Japan)
[3]Eastern Hokkaido Wildlife Station, Institute of Environmental Science, Hokkaido Research Organization, Kushiro, Hokkaido, Japan
[4]Wild Animal Medical Center, Graduate School of Veterinary Medicine, Rakuno Gakuen University, Ebetsu, Hokkaido, Japan

Abstract

Most parasitic animals including helminths, e. g., trematodes, cestodes, nematodes, acanthocephalans, leeches etc, and arthropods, e. g., ticks, mites and tongue worms, are component parts of the ecosystems in which their wild avian hosts occur, they have co-evolved, and the infectious agents do not necessarily cause disease. Sometimes, however, some individuals become sick and in extreme cases die, in various parts of the world

* Correspondence author: askam@rakuno.ac.jp.

because of an outbreak of an epidemic disease. And, considerable environmental change has occurred in Japan, especially, most northern part of Hokkaido Island, over the last 150 years as well. Furthermore, the island is one of important transit sites of the country for a wide range of migratory avian species, because that is situated on the East Asian Flyway a migratory route. Hence, the mortality or subclinical diseases due to the parasitic animals mentioned above have not so far significantly impacted populations of birds only in Japan. Because an understanding of the parasitic avian disease organisms that are present in Japan, their ecology, transmission, and symptoms is an essential conservation tool, we have overviewed this field by the end of the 20 century. However, according to the decade passed by, several issues have to be added to the review paper. So, Yoshino and Asakawa (2013) presented the nematode diseases and/or agents recorded from anatid species on the island. Hence, this chapter provides a recent overview of the parasitic diseases and/or their responsible agents recorded on non-anatid avian species as the succession, and briefly potential strategies for epidemic risk reduction are given based on the data from Wild Animal Medical Center, Rakuno Gakuen University. By the present, parasitic helminths including total 175 trematode, 148 cestode, 70 nematode, and 20 acanthocephalan species have been recorded. Among trematodes, *Leucochloridium* sp., *Renicola* sp., *Balfouria monogama* etc were found. Several cases of nematodiasis including fatal cases were reported, for example, eustrongylidosis was confirmed as one of factors contributing to the mortality of such piscivorous birds in Japan, too. The spirurid belonging to the genera *Thelazia*, *Acuaria, Cosmocephalus, Dispharynx, Synhimantus, Tetrameres* etc have been found among various avian species including endangered ones by our recent studies. As other nematodes, e. g., genera *Heterakis, Porrocaecum*, *Strongyloides, Hamatospiculum, Desmidocercella* etc were reported from not only intestines, but also air sac. Several acanthocephalan, bloodsucking leeches, e. g., *Theromyzon*, external/internal parasitic arthropods could affect many free-ranging birds. The parasitic diseases prevention will need to be based not only on an understanding of avian ecology but also of parasites. And, in order for medical, conservation ecological and/or veterinary medical researchers to be able to prepare for future epidemic issues amongst birds, they must perform epidemiological surveys on some highly pathogenic parasites mentioned above in wild bird populations in Japan including Hokkaido Island. In short, an international cooperation is essential in parasitic disease prevention. And, furthermore, proper educational activities for a citizen of the area are needed for the trial as well. For example, we know that parasites from a wild bird may be a popular exhibition to general citizen, because of showing indirectly an ecological aspect of the bird species. That is very important for the potential strategies against the parasitic disease control because of a holder of the casting vote for the avian future is eventually such people.

Background as Japanese Avifauna

Most parasitic animals including helminths, e.g., trematodes, cestodes, nematodes, acanthocephalans, leeches etc, and arthropods, e.g., ticks, mites and tongue worms, are component parts of the ecosystems in which their wild avian hosts occur, they have co-evolved, and the infectious agents do not necessarily cause disease. Sometimes, however, some individuals become sick and in extreme cases die, in various parts of the world because of an outbreak of an epidemic disease (Friend and Franson, 1999). And, as mentioned in another chapter of this book (Hirayama et al., in press) and Asakawa et al. (2013), considerable environmental change has occurred in Japan, especially, most northern part of

Hokkaido Island, over the last 150 years as well. Furthermore, the island is one of important transit sites of the Asian countries for a wide range of migratory avian species, because that is situated on the East Asian Flyway a migratory route, connecting North Pole, Asia, Australia, NewZealand etc (Committee for Check-List of Japanese Birds of Ornithological Society of Japan, 2012).

In short, the mortality or subclinical diseases due to the parasitic animals mentioned above have not so far significantly impacted populations of birds only in Japan.

Aim of This Chapter

Because an understanding of the parasitic avian disease organisms that are present in Japan, their ecology, transmission, and symptoms is an essential conservation tool (Friend and Franson, 1999), we have overviewed this field by the end of the 20 century (Asakawa et al., 2002). However, according to the decade passed by, several issues, especially in Hokkaido Island, have to be added to the review paper. So, Yoshino and Asakawa (2013) presented the nematode diseases and/or agents recorded from anatid species (Anseriformes) on the island. Actually, this review is mainly regarded as a part of T. Yoshino's Ph. D. works performed in the Wild Animal Medical Center (abbreviated to WAMC) among the Teaching Animal Hospital buildings belonging to the Graduate School of Veterinary Medicine, Rakuno Gakuen University (abbreviated to RGU) from 2002 to 2011. Hence, this chapter provides a recent overview of the parasitic diseases and/or their responsible agents recorded on non-anatid avian species as the succession, and briefly potential strategies for epidemic risk reduction are given.

WAMC and Challenge to a Wildlife Diseases

As mentioned by Asakawa (2010), wild animals have absolutely different aspects, namely the resources versus the issues, for human society, and the responsible bodies, mainly national and/or local governments, may learn the management of the issues one by one. For example, the Master of Science in Wild Animal Health and Biology courses organized by Zoological Society of London and the Royal Veterinary College, UK, which M. Asakawa, one of the present authors has taken, has challenged the issues. Such an educational system has been started in European countries, USA and Australia. The RGU established as an agricultural college in 1940, has 2 faculties of applied zoology and a veterinary school, and has contributed to tackle the wildlife issues. In 2004, to provide facilities for conservation medical research and education, the WAMC was established in the RGU. Principally, the main research and educational activities of the WAMC are veterinary medicine for wild and zoo animal medical practice, epidemiology on infectious pathogens of both humans and (captive and wild) animals, bio/chemical toxic agents, positive diagnosis of parasitic diseases etc. with zoo vets, university staffs, NGOs and students who are interested in the conservation medicine (Figures 1 and 2).

Because M. Asakawa who has been a wildlife helminthologist (Asakawa and Hasegawa, 2012) has managed the WAMC as a director, most scientific papers which have been published in the WAMC belong to the field of helminthology, and the Japanese Society of Zoo and Wildlife Medicine has designated WAMC as the Wildlife Helminthiasis Center (abbreviated to WHC) since 2006. The designation has helped indirectly to accelerate making solid relationships between zoological gardens and aquariums' directors/wildlife officers and researchers/exotic veterinarians all over the Japan and the WAMC. Such relationships provide not only a clinical benefit with its positive diagnosis but also educational activities as byproducts of exchanging staffs of the organizations and university students including postgraduate ones.

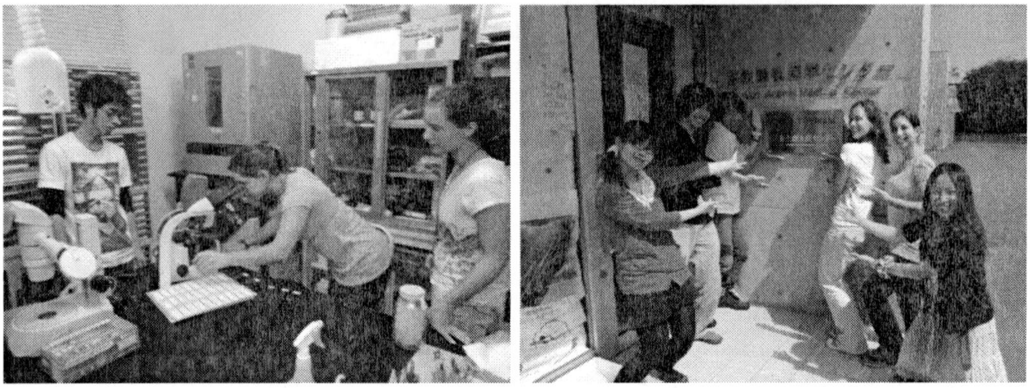

Figure 1. Wildlife medical drills including microscopic diagnosis for overseas vet students from the Ohio State University, USA (left), with Japanese ones belonging to the Wild Animal Medical Center (right) coordinated by the Extension Center, Rakuno Gakuen University, Japan (after Asakawa, 2013a).

Figure 2. Avian medical drill for overseas pre-vet students from the University of Findlay, USA, in the Wild Animal Medical Center coordinated by the Extension Center, Rakuno Gakuen University, Japan (after Asakawa, 2011).

Parasitic Helminth Diseases and Their Agents

Just after our review (Asakawa et al., 2002), we could list or mention about the parasitic helminths including total 175 trematode, 148 cestode, 70 nematode and 20 acanthocephalan species from Japanese birds (Asakawa and Hasegawa, 2003; Hasegawa and Asakawa, 2003), furthermore, add mainly a lot of nematodes to the list as mentioned later. But, anyway, it seems probable that most of these helminths including nematodes are common among Chinese, Far East Russian, Southeast Asian and/or Australian birds because in most cases their bird hosts are migratory. Moreover, many of the helminths recorded among the birds resident on Hokkaido Island may also be shared by Eurasian species because allied hosts occur in the continent as non-pathogenic commensal organisms inside bodies of the wild birds. Even the presence of these helminths in an avian species sometimes can provide us with an ecological perspective such as food habits or migration route of the bird such as the biological tag (Asakawa, 2009; Asakawa and Hasegawa, 2003; Hasegawa and Asakawa, 2003; Yoshino et al., 2012b).

Platyhelminths: Among trematodes recently found by the WAMC, *Leucochloridium* sp. (Strigeata) was obtained from the Hazel Grouse, *Bonasa bonasia vicinitas* (Ito et al., 2012), but the trematode larvae was already reported from *Succinea* sp. on Hokkaido (Asakawa and Konno, 2000). After them, such infected snails (Figure 3) have often been found on the island, and now, we will try to perform an experimental infection to chicks with the larvae (Ushiyama et al., unpubl.). Adding to this, Asakawa et al. (2008) could observe *Renicola* sp. (Plagiorchiida) from a histopathological specimen derived from one individual of mass mortality of *Puffinus tenuirostris* at several sea coasts of Hokkaido, and this case was due to concomitant infection with a renal coccidian protozoa (probably, *Eimeria* sp.) and bacteria (*Corynebacterium* sp.) (Figure 4). As one of tasks offered to the WHC, we could experience a severe enteritis caused by *Balfouria monogama* (Echinstomida) from a captive Marabou Stork, *Leptoptilos crumeniferus* kept in a zoological garden in Yokohama, near to Tokyo (Yoshino et al., 2011). Adding to this, a similar lesion was found from a captive Oriental White Stork (*Ciconia boyciana*) in the Japanese Stork Center performing re-introduction to the Japanese ecosystem (Asakawa, unpubl.). Because 2nd intermediate hosts of the echinostomid are fresh water fishes, we have to perform presence or absence check about the echinostomid in endangered piscivorous birds including the storks and Asian Crested Ibises (*Nipponia nippon*), too.

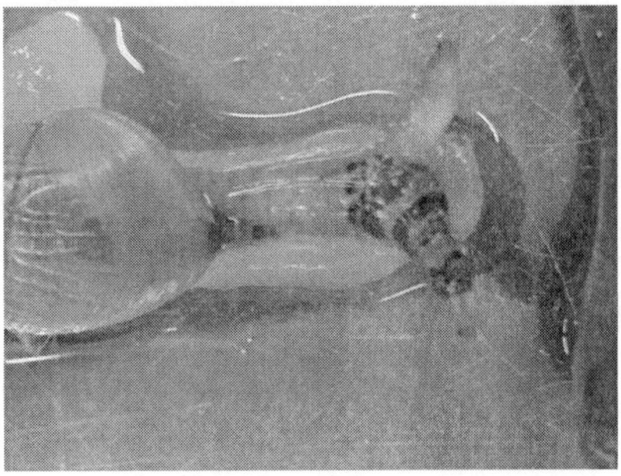

Figure 3. Land snail, *Succinea* sp., parasitized by a sporocyst of *Leucochloridium* sp. on Hokkaido Island. (after Ushiyama, unpubl.).

Nematodes: During between 1990 and 2000, several cases of helminthiasis, including fatal cases, were reported among wild waterfowl in Japan. For example, *Eustrongylides tubifex* (Family Dioctphymatidae) is commonly found among piscivorous water birds as not only final hosts (Friend and Franson, 1999; Asakawa et al., 2002; Figure 5), but also play paratenic hosts (Yoshino and Asakawa, 2012). Recently, it has been speculated that there is an intimate relationship between the nematode infection and water pollution (eutrophication; Asakawa, 2013b). Anyway, it is well-known that eustrongylidosis is one of factors contributing to the mortality of such piscivorous birds (Friend and Franson, 1999).

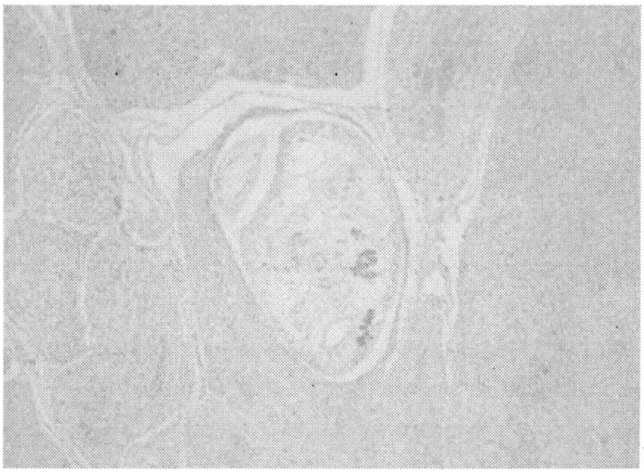

Figure 4. A histopathological specimen of a kidney parasitized by *Renicola* sp., from the shearwater (*Puffinus tenuirostris*) at sea coast of Hokkaido Island (after Asakawa et al., 2008).

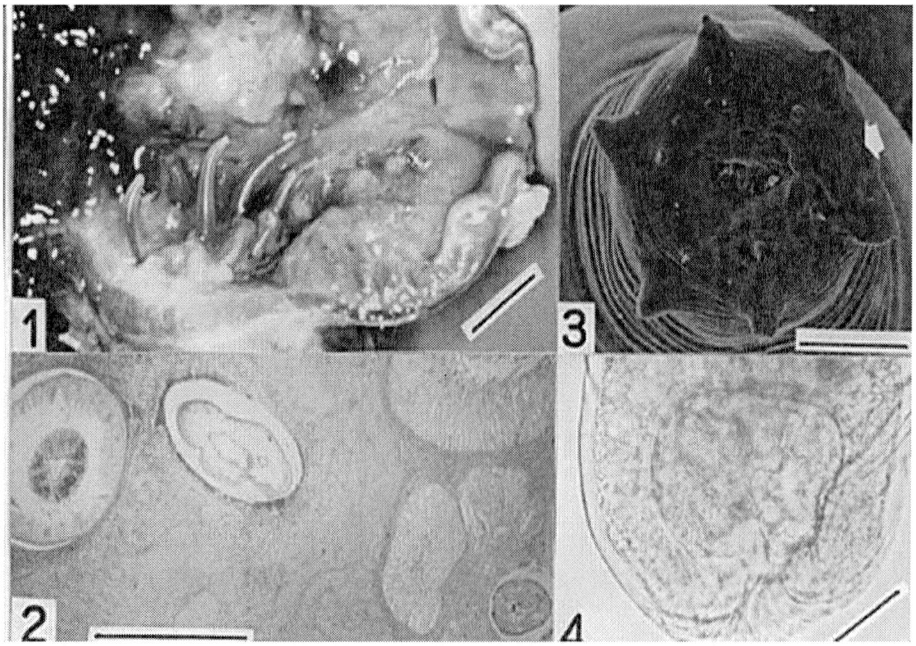

Figure 5. Fatal eustrongylidosis and *Eustrongylides tubifex* (responsible agent) in an immature wild little grebe (*Tachybaptus ruficollis*) from Japan (-1: Its gross lesion, -2: Histopathiological lesion, -3: Apical view of the nematode, -4: Tail end of the female nematode; after Asakawa, 2013).

The spirurid genus *Thelazia* parasitizes the eye (conjunctivae, nictitating membrane and lacrimal duct) and can cause conjunctivitis, and Asakawa et al. (2002) mentioned that *T.* (*Thelaziella*) *aquillina* was obtained from a captive and free-ranging the Oriental White Stork after the re-introduction. Since this bird-specific subgenus *Thelaziella* is well known as a pathogen causing conjunctivitis, an epidemic program of preventing infection by the nematode at the center have established in order to provide adequate conditions for the conservation of the captive storks. The other spirurid genera also include gastrointestinal parasites such as *Acuaria, Cosmocephalus, Dispharynx, Synhimantus, Tetrameres* etc, which can cause ulceration (Figure 6), hemorrhage, and proliferative inflammatory nodular lesions on the mucosal surface of the gastrointestinal tract. Such nematodes have been found among various avian species by our recent studies including the Red-crowned Cranes (*Grus japonensis*; Figure 7), (Nakamura et al., 2004; Sato et al., 2005; Ohshima et al., 2005; Onuma et al., 2008; Saito et al., 2006; Watanabe et al., 2006; Yoshino et al., 2005a, b, 2009a, b, 2013; Uematsu et al., 2013; Hirayama et al., 2013).

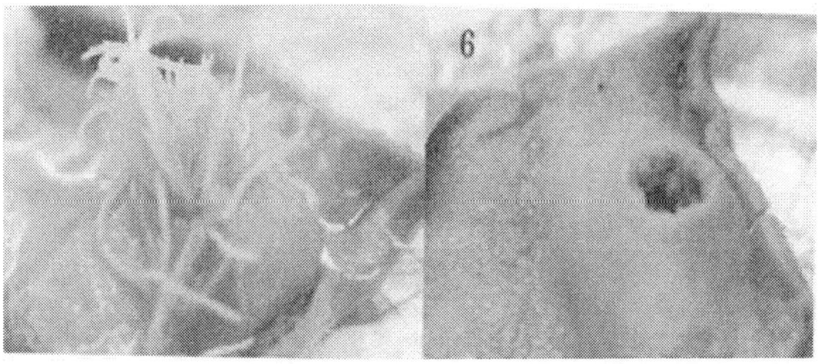

Figure 6. Ulceration on avian stomach mucosa due to nematode parasitization (after Yoshino, unpubl.).

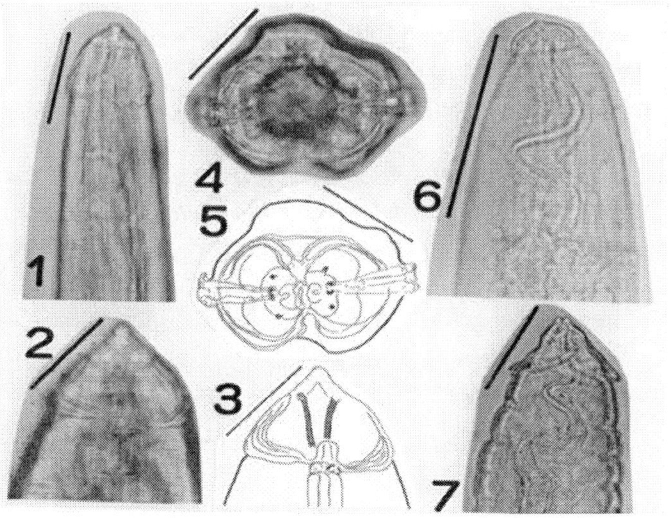

Figure 7. Spirurid genera, e. g., *Paracuaria* (from -1 to -3), *Chordocephalus* (from -4 to -6) and *Viktrocara* (-7), from Red-crowned Cranes in Hokkaido Island. (after Ohshima et al., 2005, figures in submission).

The superfamily Trichostrongyloidea including genera *Trichostrongylus, Amidostomum* and *Ornithostrongylus* is common avian gastrointestinal nematode group as well, and the later 2 genera *Amidostomum* and *Ornithostrongylus* were recorded in Japan (Asakawa and Hasegawa, 2003). Although the genus *Trichostrongylus* has a wide range of mammalian hosts, e.g., cattle, sheep, rabbits, rodents etc (Hasegawa and Asakawa, 2003; Asakawa and Hasegawa, 2012), there is no record of detection of the genus *Trichostrongylus* from the Japanese birds. It is well known, for example, that *T. tenuis* can even serve to regulate the populati on of Red Grouse (*Lagopus lagopus*) in the United Kingdom causing reduced fecundity and survival among wild grouse (Friend and Franson, 1999; Asakawa et al., 2002), however, the genus *Trichostrongylus* has not yet been recorded from the Japanese *Lagopus* (Ito et al., 2012; Yoshino et al., 2008b).

In stead of the superfamily, another intestinal nematode genus *Heterakis* (Heterakoidea) has been newly recorded from both wild and zoo captive birds (Ushigome et al., 2010, 2011; Yoshino et al., 2008b; Onuma et al., 2011b). Among the species belonging to the nematode genus newly found by the WAMC, it was prominent that *H. isolonche* was obtained from endangered species, Okinawa rails (*Gallirallus okinawae*) (Yoshino et al., 2008b). From the conservational point of view, a nested PCR assay for molecular diagnosis and identification of nematode has been performed for health monitoring for the capitive rails (Zhao et al., 2009; Onuma et al., 2012; Onuma and Zhao, 2013).

The other gastrointestinal nematode group including the genus *Porrocaecum* (Ascaridoidea), *Strongyloides* etc (Rhabditoidea), and the respiratory (air sac) ones including *Hamatospiculum* (Diplotriaenoidea) and *Desmidocercella* (Aproctoidea) were reported by the WAMC, respectively (Onuma et al., 2006, 2011a; Uematsu et al., 2012; Yoshino et al., 2005b, 2008a, 2012a, e).

Acanthocephalans: Several acanthocephalan species were newly reported from avian species including alien ones to Japan by the WAMC (Nakamura et al., 2004; Yoshino et al., 2003, 2009a, b; Sato et al., 2005). Because some of the acanthocephalans regarded as responsible agents for severe enteritis, that surveillance should be kept. By the way, we have an interesting case about the materials derived from conservation ecologists to the WAMC that over 10 individuals of *Polymorphus* sp. which seemed to be "extracted" from aquatic birds (?) were found in dredging samples derived from Lake Biwa (Asakawa and Nishino, 2011).

Leeches: Bloodsucking leeches (families Hirudinidae and Haemadipsidae) including the genera *Hirudo, Myxobdella, Macrobdella, Ornithobdella, Hirudobdella, Haemadipsa, Chtonobdella* and *Planobdella*, *Theromyzon* etc affect many aquatic bird species (Asakawa et al., 2002). Among them, there have been some reports of disease caused by these nasal or orbit leeches belonging to the genus *Theromyzon* from water birds (Suda et al., 2006; Yoshino et al., unpubl.) in Hokkaido.

Parasitic Arthropod Diseases and Their Agents

Infestations of avian external parasitic arthropods, namely ticks, mites, bugs, fleas, flies etc such as the families and genera shown in the table 1, directly cause or increase the likelihood of anemia, feather damage, nestling mortality, and nest desertion.

Table 1. Parasitic arthropods from birds in the whole world (after Aasakawa et al., 2002)

Class Arachnida, Subclass Acari (ticks and mite)
Order Parasitiformes
Suborder Metastigmata (ticks, larvae-adult)
Family Argasidae (soft ticks): *Argas, Ornithodoros*
Ixodidae (hard ticks): *Amblyomma, Ixodes, Haemaphysalis*
Suborder Mesostigmata
Family Rhinonyssidae (respiratory [nasal] mites): *Haemolaelaps, Rhinonyssus, Mesonyssus* etc
Dermanyssidae (nest mite, nymph & adult): *Dermanyssus*
Macronyssidae (ditto): *Ornithonyssus*
Laelapidae (ditto) *Mesonyssoides*
Order Acariformes
Suborder Prostigmata
Family Trombiculidae (chigger mite, larvae): *Trombicula, Leptotrombidium, Mackiena* etc
Harpyrhynchidae (skin mites, under cuticle): *Harpyrhynchus*
Cheyletidae (skin mites, construct silken'nests' on skin): *Cheyletus, Bakericheyla* etc
Syringophilidae (quill mites): *Syringophilus*
Ereynetidae (nasal cavity): *Trispeleognathus, Neoboydaia*
Cloacaridae (lung): *Epimydodex*
Suborder Astigmata
Family Hypoderatidae (subcutaneous mites, nymphs): *Phalacrodectes, Hypodectes*
Laminosioptidae (subcutaneous and quill wall mites): *Laminosioptes*
Knemidocoptidae (skin [scaly leg/face and depluming] mites): *Knemidocoptes, Procnemidocoptes,Neocnemidokoptes*
Cytoditidae (respiratory [air sac] mites): *Cytodites*
Turbinoptidae (respiratory [nasal] mites): *Turbinoptes, Rhinoptes*
Epidermoptidae (skin mites, embedded in and under skin or in feather bulbs): *Epidermoptes, Myialges, Microlichus*
Dermationidae (ditto): *Dermation, Neodermation, Paddacoptes, Passeroptes* etc
Freyanidae (feather [outside] mites): *Freyana*
Analdidae (ditto): *Analges, Megninia, Bychovskiata*
Pterolichidae (ditto): *Pterolichus, Ascetolichus, Aralichus*
Alloptidae (ditto): *Alloptes*
Proctophyllodidae (ditto): *Proctophyllodes, Trouessartia*
and other Fams. Avenzoariidae, Trouessartiidae, Falculiferidae, Kramerellidae
Dermoglyphidae (quill mites, or feather [inside] mites): *Dermoglyphus, Falculifer*
and other Fams. Apionacaridae, Gaudoglyphidae, Syringobiidae

Class Insecta
Order Mallophaga (feather, biting, chewing lice)
Suborder Amblycera
Family Menoponidae: *Menacanthus, Menopon, Trinoton* etc
Laemobothriidae: *Laemobothrion*
Ricinidae: *Ricinus*
Suborder Ischnocera
Family Philopteridae: *Acidoproctus, Anaticola, Anatoecus, Columbicola, Craspedonirmus, Cuculoecus, Lagopoecus, Lipeurus, Strigiphilus, Sturnidoecus* etc
Heptapsogasteridae: *Heptapsogaster*
Order Hemiptera (true bugs)
Family Cimicidae: *Oeciacus, Cimex, Haematosiphon* etc

Table 1. (Continued)

Order Siphonaptera (fleas)
Family Pulicidae: *Pulex*
Ceratophyllidae: *Orneacus, Ceratophyllus, Dasypsyllus* etc
Leptopsyllidae: *Leptopsylla, Frontopsylla, Callopsylla*
Pulicidae: *Ornithopsylla, Acetenopsylla, Echidnophaga, Xenopsylla*
Pygiopsyllidae: *Stivalius, Notiopsylla*
Rhopalopsyllidae: *Listronius, Parapsyllus*
Tungidae: *Hectopsylla*
Order Diptera (flies)
Family Calliphoridae (blowflies): *Trypocalliphora, Protocalliphora*
Hippoboscidae (louse flies): *Ornithoica, Ornithomya, Crataerina, Ornithoctona*
Carnidae (milichiid flies): *Carnus*
Muscidae (botflies): *Philornis*
Sarcophagidae (flesh flies): *Wohlfahrtia*
Neottiophilidae (neottiophid flies): *Neottiophilum*
and other Fams. Culicidae (mosquitoes), Simuliidae (blackflies), Ceratopogonidae (biting midges), and Tabanidae (horseflies and deerflies) (feeding on blood during ephemeral visiting)

Class Crustacea
Order Pentastomida (tongue worms)
Family Reighardiidae: *Reighardia*

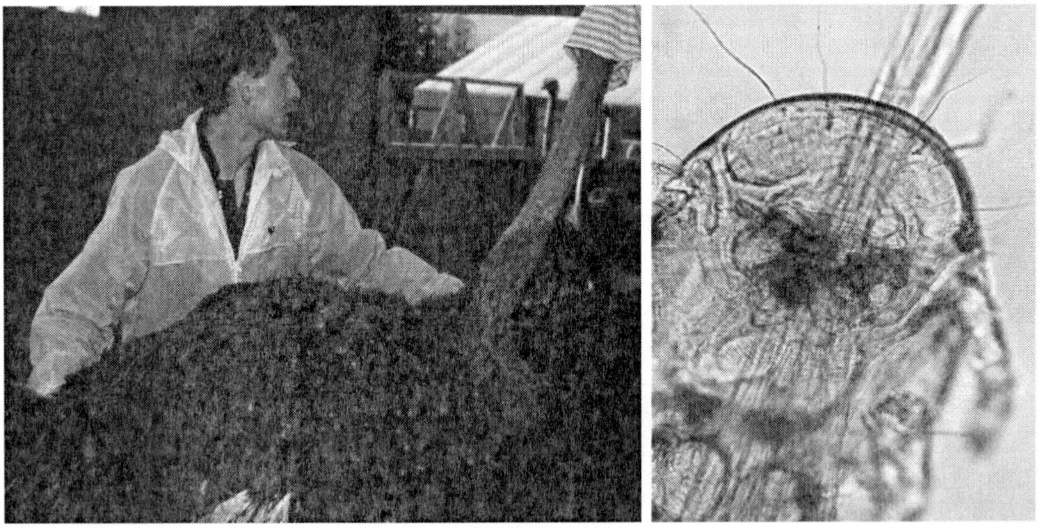

Figure 8. A captive female ostrich kept in a ostrich farms in Hokkaido (left), parasitized by feather lice (*Struthiolipeurus struthionis;*right; after Asakawa, 2000, 2003).

Recently, several cases were reported from avian species including alien ones to Japan by the WAMC (Yoshino et al., 2003, 2009b, 2012c; d; Uematsu et al., 2013). Adding to these free-ranging birds, ostrich feather lice (*Struthiolipeurus struthionis*) from captive ostrich in a farm of Hokkaido (Asakawa, 2000, 2003; about the ostrich farm, see, Asakawa, 2010; (Figure 8).

Their indirect effects include the transmission by some species of microbial, protozoan or helminth infections, and, according to the another chapter inside this book (Hirayama et al., in press), there have been reported cases or epidemic surveys of the diseases in relation to the avian parasitic arthropods in Japan. On the other side, it is well known that some of arthropods are internal parasites, e. g., respiratory or nasal mites, subcutaneous mites, tongue worms etc, and several cases were reported by the WAMC (Nakamura et al., 2003; Uemura et al., 2010).

Conclusion

Diseases prevention will need to be based not only on an understanding of avian ecology but also of parasites. And, in order for medical, conservation ecological and/or veterinary medical researchers to be able to prepare for future epidemic issues amongst birds, they must perform epidemiological surveys of some highly pathogenic parasites mentioned above in wild bird populations in Japan including Hokkaido Island and its surrounding areas. In short, an international cooperation is essential in parasitic disease prevention. And, furthermore, proper educational activities for a citizen of the area are needed for the trial as well. For example, we know that parasites from wild a bird may be a popular exhibition to general citizen, because of showing indirectly an ecological aspect of the bird species (Aizawa et al., 2006). That is very important for the potential strategies against the parasitic disease control for a holder of the casting vote for the avian future is eventually such people.

Acknowledgments

The present chapter was supported in part by the priority research (2502-11E004) of Hokkaido Research Organization, and by the Supported Program for the Strategic Research Foundation at Private Universities (2013) of the Ministry of Education, Culture, Science and Technology, Japan, respectively.

References

Aizawa, K., Yoshino, T., Sawahata, T., Nagano, M. and Asakawa, M. 2006. Host-parasite relationships as natural historical exhibition: A case of avian parasite fauna including helminthes and mites in Matsunoyama District and its adjacent areas, Niigata, Prefecture, Japan. *The Proceedings Joint International Symposium of the12th Annual Meeting of Japanese Society of Zoo and Wildlife Medicine and the 21st Century COE Program of Gifu University*, Japan: 179. (in Japanese with English summary).

Asakawa, M. 2000. [(translated title) Preliminary parasitological investigation at an ostrich farm in Hokkaido, Japan]. *J. Hokkaido Vet. Med. Assoc.*, 44: 296-299 (in Japanese).

Asakawa, M. 2003. [(translated title) Parasites recorded captive and/or wild ostriches, with special references to Japanese cases]. *J. Ostrich and Ratite Sci.*, (3): 19-25. (in Japanese).

Asakawa, M. 2009. [(translated title) Comment on the food habit of *Otus scops* deduced from acanthocephalns]. *Shinrin-hogo*, (314): 13-14. (in Japanese).

Asakawa, M. 2010. Ecotourism with utilization of wild animals - Its impact on conservation medicine and risk assessment in Hokkaido, Japan. *In*: (Krause, A. and Weir, E. Eds.) *Ecotourism: Management, Development and Impact*, Nova Science Publishers, Inc., NY, USA: 227-240.

Asakawa, M. 2011. [(translated title) Annual report 2010 of the Wild Animal Medical Center, Rakuno Gakuen University (3).]. *J. Hokkaido Vet. Med. Assoc.*, 55, 139-144. (in Japanese).

Asakawa, M. 2013a. [(translated title) Annual report 2012 of the Wild Animal Medical Center, Rakuno Gakuen University (3)]. *J. Hokkaido Vet. Med. Assoc.*, 57: 95-99. (in Japanese).

Asakawa, M. 2013b. [(translated title) Avian helminths and water environmental pollution]. *J. Jpn. Vet. Med. Assoc.*, 66: 366-367. (in Japanese).

Asakawa, M. and Hasegawa, H. 2003. Parasitic nematodes from avian and mammalian species in Japan. Bull. biogeogr. Soc. Japan, 58: 79-93. (in Japanese with English summary).

Asakawa, M. and Hasegawa, H. 2012. Conservation medical strategy of endemic host-parasite relationships- Case model of Japanese rodents (Mammalia) and their nematodes. *In*: (Boari, F. and Chung, J. A. Eds.) *Nematodes: Morphology, Functions and Management Strategies*. Nova Science Publishers, NY, USA: 239-249.

Asakawa, M., Horikami, A., Wada, M., Aizawa, K., Watanabe, H., Yoshino, T., Okamoto, M., Onuma, M., Murata, K.and Kuwana, T. 2008. Post mortem examination of mass death carcasses of Short-tailed Shearwater (*Puffinus tenuirostris*) found in northern Hokkaido, Japan, with special reference to parasitological observation. *Jpn. J. Vet. Parasitol.*, 7 (1): 21. (in Japanese).

Asakawa, M. and Konno, T. 2000. [(translated title) A land snail parasitided by *Leucochloridium* sp.] *J. Hokkaido Vet. Med. Assoc.*, 44: 411. (in Japanese).

Asakawa, M., Nakade, T., Murata, S., Ohashi, K., Osa, Y. and Taniyama, H. 2013. Recent viral diseases of Japanese anatid with a fatal case of Marek's disease in an endangered species, white-fronted goose (*Anser albifrons*). *In:* (Hambrick, J. and Gammon, L. T. Eds.) *Ducks: Habitat, Behavior and Diseases*, Nova Science Publishers, NY, USA: 37-48.

Asakawa, M., Nakamura, S. and Brazil, M. A. 2002. An overview of infectious and parasitic diseases in relation to the conservation biology of the Japanese avifauna. *J. Yamashina Inst. Ornithol.*, 34: 200-221.

Asakawa, M. and Nishino, M. 2011. A case report of acanthocephalans found in dredging samples derived from Lake Biwa, Japan. *J. Rakuno Gakuen Univ., Nat. Sci.*, 35: 91-94. (in Japanese with English summary).

Committee for Check-List of Japanese Birds of Ornithological Society of Japan (Ed.). 2012. *Check-List of Japanese Birds,* 7th ed. Ornithological Society of Japan, Tokyo, Japan: 438 pp.

Friend, M. and Franson, J. C. (Eds.) 1999. *Field Manual of Wildlife Diseases*, USGS, Washington D. C., USA: 426 pp.

Hasegawa, H. and Asakawa, M. 2003. Parasitic helminth fauna of terrestrial vertebrates in Japan. *In*: (Otsuru, M., Kamegai, S. and Hayashi, S. Eds.) *Progress of Medical Parasitology in Japan*, Vol. 7, Meguro Parasitological Museum, Tokyo, Japan: 129-145.

Hirayama, T., Chiba, T., Sugawara, R., Yamada, N., Osa, Y. and Asakawa, M. 2013. Parasitological examination performed on rescued wild birds: its preliminary results and necessity. *The Proceedings of 19th Annual Meeting of Japanese Society of Zoo and Wildlife Medicine*, Kyoto University of Kobe, Japan: in press. (in Japanese).

Hirayama, T., Ushiyama, K., Osa, Y. and Asakawa, M. in press. Recent infectious diseases or their responsible agents recorded from Japanese wild birds. *In*: *Birds: Evolution and Behavior, Breeding Strategies, Migration and Spread of Disease*, Nova Science Publishers, NY, USA: in press.

Ito, H., Yoshino, T., Nakamura, S., Endo, D., Fijimaki, Y., Nakada, T. and Asakawa, M. 2012. Isolation of three nematode species from the Hazel Grouse, *Bonasa bonasia vicinitas* Riley, 1915, in Hokkaido, Japan. *Jpn J. Zoo Wildl Med.* 17: 21-25.

Nakamura, S., Morita, T. and Asakawa, M. 2003. New host records of arthropod parasites from sea birds in Hokkaido, Japan. *Jpn. J. Zoo Wildl. Med.* 8(2): 131-133.

Nakamura, S., Yoshino, T., Sato, J., Chiba, A. and Asakawa, M. 2004. The parasitic helminths from avian species in Niigata Pref., Japan. *Jpn. J. Ornithol.*, 52: 116-118. (in Japanese with English summary).

Ohshima, Y., Uebayashi, A., Shimura, R., Osa, Y., Taniyama, H. and Asakawa, M. 2005. Helminthological survey on Japanese cranes (*Grus japonensis*) in Hokkaido, Japan. *The Proceedings of 11th Annual Meeting of Japanese Society of Zoo and Wildlife Medicine*, Obihiro University of Agriculture and Veterinary Medicine, Japan: 93. (in Japanese).

Onuma, M., Yoshino, T., Mizuo, A., Kakogawa, M., Asakawa, M. 2011a. First host record of *Porrocaecum semiteres* (Zeder, 1800) Baylis, 1920 (Nematoda: Ascaridoidea) obtained from a Superb Starling, *Lamprotornis superbus* Ruppell, 1845 with an overview of the genus *Porrocaecum* recorded from Japanese birds. *Biogeography,* 13: 59-63.

Onuma, M., Yoshino, T., Takagi, M., Matsui, S., Hisaka, M., Nagamine, T., Murata, K., Kuwana, T. and Asakawa, M. 2006. A case report of the genus *Hamatospiculum* (Diplotriaenoidea: Nematoda) obtained from Bull-headed Shrikes (*Lanius bucephalus*) and its conservation medical point of view. *J. Vet. Parasitol.*, 5 (1): 50. (in Japanese).

Onuma, M., Yoshino, T., Watanabe, H., Ohshima, Y., Okamoto, M., Shimura, R., Watanabe, Y., Saito, K., Kuwana, T., Murata, K., Taniyama, H. and Asakawa, M. 2008. An overview of epidemiological investigations of helminth parasites for endangered avian species in Japan. *J. Vet. Parasitol.*, 6(2): 86. (in Japanese).

Onuma, M., Yoshino, T., Zhao, C., Nagamine, T., Asakawa, M. 2011b. Parasitic helminths obtained from Okinawa rails, *Gallirallus okinawae*. *J. Yamashina Inst. Ornithol.*, 43: 74-81.

Onuma, M. and Zhao, C. 2013. Molecular diagnosis of *Heterakis isolonche* parasitic in Okinawa rails (*Gallirallus okinawae*). In: (Asakawa, M. Ed.) *Materials and Methods in Parasitology*, Sankei-sha Co.Ltd, Nagoya, Japan: in press. (in Japanese with English summary).

Onuma, M., Zhao, C., Asakawa, M., Nagamine, T. and Kuwana, T.2012. Duplex real-time PCR assay for the detection of two intestinal parasites, *Heterakis isolonche* and *Glaphyrostomum* sp., in Okinawa rail (*Gallirallus okinawae*) *Jpn. J. Zoo Wildl. Med.*, 17: 27-31.

Saito, K., Yoshino, T., Aizawa, K., Kotaka, N., Ueno, M. and Asakawa, M. 2006. Parasitic nematodes found from injured Ruddy Kingfisher (*Halcyon coromanda bangsi*) in Okinawa Prefecture, Japan, with special reference to a clinical case report of stomatitis infested with acuarid nematodes. *The Proceedings Joint International Symposium of the12th Annual Meeting of Japanese Society of Zoo and Wildlife Medicine and the 21st Century COE Program of Gifu University, Japan*: 183. (in Japanese with English summary).

Sato, A., Nakamura, S., Takeda, M., Murata, K., Mitsuhashi, Y., Kawai, N., Tanaka, N. and Asakawa, M. 2005. Parasitic helminths from exhibited avian species kept in Kinki District in Japan. *Jpn. J. Zoo Wildl. Med.* 10 (1): 35-38.

Suda, S., Endo, D., Osa, Y. and Asaakwa, M. 2006. Taxonomic study on a duck leech, *Theromyzon* sp. (Rhynchobdellida: Glossiphoniidae) from wild geese in the Kamchatka Peninsula, Russia. *Jpn. Soc. Med. Entomol. Zool.*, 57: 152. (in Japanese).

Uematsu, J., Kanesaka, H. and Asakawa, M. 2012. New local record of parasitic nematode *Desmidocercella incognita* from a cormorant (*Phalacrocorax carbo*) in Japan. *Proc. Jpn. Assoc. Clin. Avian Med.,* (15): 15-16. (in Japanese).

Uematsu, J., Tsuchiya K. and Asakawa, M. 2013. Application for an epidemiological survey used with carcasses derived from bird control performing on international air ports in Japan with a special reference to preliminary results of parasitological research. *J. Rakuno Gakuen Univ., Nat. Sci.*, 37: 167-172. (in Japanese with English summary).

Uemura, J., Kinbara, R., Yoshino, T., Morita, T., Kataoka, A., Taniyama, H., Fukui, D., Endoh, D. and Asakawa, M.2010. [(translated title) Several case reports of dermatid mites from *Coccothraustes coccothraustes*]. *Jpn. Assoc. Clin. Avian Med.*, (13): 48-50. (in Japanese).

Ushigome, N., Yoshino, T., Suzuki, Y., Kawajiri, M., Masaki, K., Endo D. and Asakawa, M. 2010. Three species of the genus *Heterakis* (Nematoda: Heterakidae) from a captive Satyr Tragopan (*Tragopan satyra*) (Aves) in a zoological garden. *Nematol. Res.,* 40: 21-23.

Ushigome, N., Yoshino, T., Suzuki, Y., Kawajiri, M., Masaki, K., Endo, D. and Asakawa, M. 2011. The parasitological survey on animals in Kawasaki Yumemigasaki Zoological Park. Jpn. *J. Zoo Wildl. Med.,* 16 (2): 133-137. (in Japanese with English summary)

Watanabe, H., Yoshino, T., Osa, Y., Okamoto, M. and Asakawa, M. 2006. Epidemiological and pathological examinations of parasitic helminths found from endangered and vulnerable sea bird species belonging to the family Alcidae in Hokkaido, Japan. *The Proceedings Joint International Symposium of the12th Annual Meeting of Japanese Society of Zoo and Wildlife Medicine and the 21st Century COE Program of Gifu University, Japan*: 177. (in Japanese with English summary).

Yoshino, T. and Asakawa, M. 2012. [(translated title) A hepatic eustrongylidosis due to parasitization by its larva in a little grebe]. *Shinrin-hogo*, (325): 5-7. (in Japanese).

Yoshino, T and Asakawa, M. 2013. A brief overview of parasitic nematodes recorded from waterfowls on Hokkaido, Japan. *In*: (Hambrick, J. and Gammon, L. T. Eds.) *Ducks: Habitat, Behavior and Diseases*, Nova Science Publishers, NY, USA: 59-64.

Yoshino, T., Azumano, A., Endo, D. and Asakawa, M. 2011. Morphological and pathological notes on *Balfouria monogama* Leiper, 1908 (Trematoda: Echinstomida) of a Marabou Stork, *Leptoptilos crumeniferus*. *J. Vet. Med.* (Tokyo), 64: 133-136. (in Japanese with English summary).

Yoshino, T., Hayakawa, D., Yoshizawa, M., Osa, Y. and Asakawa, M. 2012a. First record of *Strongyloides avium* Cram, 1929 (Nematoda: Rhabditoidea) obtained from a Fairy Pitta, *Pitta brachyura nympha* Temminck & Schlegel, 1850, kept in zoological garden. *Bull. Tokushima Pref. Mus.* (22): 1-6.

Yoshino,T., Hama, N., Yanai,T. and Asakawa, M. 2008a. [(translated title) The first report of the *Hamatospiculum* spp. nematode in Flammulated Owl (*Outs flammeolus*) and summary of the Diplotriaenoidea superfamily and Aproctoidea superfamily nematode recorded in Japan]. *The Proceedings of 14th Annual Meeting of Japanese Society of Zoo and Wildlife Medicine*, University of Kobe, Japan: 97. (in Japanese).

Yoshino, T., Kawakami, K., Sasaki, H., Miyamoto, K. and Asakawa, M. 2003. A parasitological survey of Hwamei *Garrulax canorus* and Red-billed Leiothrix *Leiothrix lutea* (Passeriforms: Terimiidae). *Jpn. J. Ornithol.*, 51(2): 39-42. (in Japanese with English summary).

Yoshino, T., Kimura, M. and Asakawa, M. 2012b. [(translated title) A cestode record from *Picus canus*]. *Shinrin-hogo*, (316): 14-16. (in Japanese).

Yoshino, T., Nakada, T. and Asakawa, M. 2012c. [(translated title) A case report on postmortem of *Jynx torquilla* with special reference to a parasitization by a fela, *Ceratophyllus tribulus*]. *J. Hokkaido Vet. Assoc.*, 56: 45-47. (in Japanese).

Yoshino, T., Nakamura S, Endoh, D., Onuma, M., Osa, Y., Teraoka, H., Kuwana, T. and Asakawa, M. 2009a. A helminthological survey of four families of waterfowl (Ardeidae, Rallidae, Scolopacidae and Phalaropodidae) in Hokkaido, Japan. *J. Yamashina Inst. Ornithol.*, 41: 42-54.

Yoshino, T., Onuma, T., Nagamine, T. and Asakawa, M. 2013. First record of *Synhimantus (Dispharynx) nasuta* (Nematoda: Acuarioidea) and *Neopteronyssus* sp. (Acari: Pteronssidae) from Okinawa woodpecker, *Sapheopipo noguchii. Jpn. Wildl. Res. Soc.,* (38): 7-12. (in Japanese with English summary).

Yoshino, T., Onuma, M., Nagamine, T., Inaba, M., Kawashima, T., Murata, K., Kawakami, K., Kuwana, T. and Asakawa M. 2008b. First record of the genus *Heterakis* (Nematoda: Heterakidae) obtained from two scarce avian species, Japanese rock ptarmigan (*Lagopus mutus japonicus*) and Okinawa rails (*Gallirallus okinawae*) in Japan. *Jpn. J. Nematol.*, 38: 89-92.

Yoshino, T., Shingaki, T., Onuma, M., Kinjo, T., Yanai, T., Fukushi, H., Kuwana, T. and Asakawa M. 2009b. Parasitic helminths and arthropods of the Crested Serpent Eagle, *Spilornis cheela perplexus* Swann, 1922 from the Yaeyama. *J. Yamashina Inst. Ornithol.*, 41: 55-61.

Yoshino, T., Shingaki, T., Yanai, T., Fukushi, H., Kinjo, T., Saito, K., Watanabe,Y., Kurosawa, N., Osa,Y. and Asakawa, M. 2005a. External and internal parasites obtained from some endangered raptor species in Japan. *The Proceedings of 11th Annual Meeting of Japanese Society of Zoo and Wildlife Medicine*, Obihiro University of Agriculture and Veterinary Medicine, Japan: 89. (in Japanese).

Yoshino, T., Uemura, J. and Asakawa, M. 2012d. Two parasite species of arthropods found in a nest and the nestling of Pied Wagtail, *Motacilla alba* Linnaeus, 1758, in Hokkaido, Japan. *J. Rakuno Gakuen Univ., Nat. Sci.,* 36 (2): 313-315.

Yoshino, T., Uemura, J., Uematsu, K., Tsuchida, S., Osa, Y., Taniyama, H. and Asakawa, M. 2005b. Necropsy and parasitological survey of seabirds killed in the oil spill at Ishikari. *The Proceedings of 11th Annual Meeting of Japanese Society of Zoo and Wildlife Medicine*, Obihiro University of Agriculture and Veterinary Medicine, Japan: 126. (in Japanese).

Yoshino, T., Yanai, T., Asano, M. and Asakawa, M. 2012e. First record of *Porrocaecum depressum* (Nematoda: Ascaridoidea), *Craspedorrhynchus* sp. and *Degeeriella* sp. (Insecta: Psocodea) obtained from a Hodgson's Hawk Eagle, *Spizaetus nipalensis,* in Gifu Prefecture, Japan. *Biogeography*, 14: 143-148.

Zhao, C., Onuma M, Asakawa M. and Kuwana T. 2009. Preliminary studies on developing a nested PCR assay for molecular diagnosis and identification of nematode (*Heterakis isolonche*) and trematode (*Glaphyrostomum* sp.) in Okinawa rail (*Gallirallus okinawae*). *Vet Parasitol.* 163: 156-160.

In: Birds
Editors: Lucas Ruiz and Franco Iglesias

ISBN: 978-1-62948-104-3

Chapter 8

Environmental Factors that Affect to Urban Avian Communities

Daniel Patón*
Numerical Ecology Unit. Faculty of Sciences,
University of Extremadura, Badajoz, Spain

Abstract

In the present study we determine in to x35 areas of Iberian Peninsula along a gradient of urbanization the following factors:After the age of the park (AG), altitude (AL), number of buildings in a circle of 500 m (BL), cover of grass (CG), shrubs (CS) and trees (CT), average noise (DB), distance to nearest city (DC), use of herbicides (HB), inhabitants of nearest city (NC), number of persons in the park (NP), density of old trees (OT), park area (PA), species richness of shrubs (RS), trees (RT) and woody plants (RW) and number of water areas (WA). We check the relationships of these environmental factors with the total bird abundance (N), the alpha-fisher diversity (α), the bird species richness (S), the Berger-Parker dominance (d), the Margalef diversity (Mg), the Shannon diversity (H), the Shannon entropy (e^h), the Simpson diversity (D) and the Simpson dominance (1-D) using stepwise multiple linear regression. To validate these statistical models, we determine the coefficients of determination (R^2), the residual normality by Shapiro-Wilk test (W) and the residual homoscedasticity by Breusch-Pagan test (BP). In all the regressions we obtained exhigh coefficients of determination, residual normality (except in Simpson dominance) and residual homoscedasticity. The main causal variables that effected ecological parameters of community were altitude, distance to nearest city, noise level and shrub cover. The park ageAge of the park was irrelevant to bebecause in urban areas the environment is continuously modified by gardening and, consequently, the succession processes are not produced. Our study shows that consequently, urban parks could be considered potential reserves for bird conservation if we manage these four factors in an optimal way.

* Corresponding author: Daniel Patón. Numerical Ecology Unit. Faculty of Sciences. University of Extremadura. Avda. Elvas s/n 06071 Badajoz, Spain.

1. INTRODUCTION

The current human population lives in urban areas and this trend has been increasing in the last few decades (United Nations Population Division, 2009). This phenomenon has a significant impact on natural ecosystems producing a high fragmentation of habitats and reducing the biodiversity (Shochat et al., 2010). In consequence, the habitat of many animal species is negatively affected. This degradation process is especially evident in very specialized species with high requirements of ecosystem size and maturity (Fernández-Juricic and Jokimäki, 2001). This simultaneous process of degradation of natural habitats and colonization of urban environments forces us to look to these urban areas as potential conservation reserves for small birds (Akinnifesi et al., 2010). Numerous studies indicate that parks and gardens in cities are places where many species of birds can potentially sustain a high density (Caula et al., 2010). It has even been shown that the abundance and variety of birds in these spaces could be higher than those of natural environments if certain conditions are maintained (Aurora et al., 2009; Patón et al., 2012). Within these features that allow the maintenance of a high representation of birds, the sensible reduction of predators characteristic of urban environments is very relevant (Rousseau, 2004).

In fact, in the Mediterranean areas of Southern Europe most urban predators are of small size such as house cats, Red Squirrels (*Sciurus vulgaris*) and Magpies (*Pica pica*), unspecialized, opportunistic, highly variable in abundance and predation intensity and easily controllable (Fernández-Juricic, 2000a). Therefore, birds in the cities live with a low rate of predation with respect to wild conditions, which undoubtedly affects their abundance positively. Another factor with a potentiating effect on the richness and abundance of birds is the high supply of food typical of urban areas. However, it is necessary to remark that this abundance of food is mainly useful for omnivorous bird species. Insectivorous birds can be adversely affected by the massive use of pesticides in urban parks and gardens that considerably reduce the abundance of insects. In this sense, the strategies of biological pest control, the increase in the use of native plants in gardens, certain organic gardening techniques such as mulching, the use of manure as fertilizer and leaving grass clippings between others are highly recommended for insectivorous bird species. In many cities, the use of artificial feeders favors small birds, especially during unfavorable seasons. Some studies have found increases in predators in the vicinity of these food dispensers, although the intensity of this adverse effect on bird populations is highly debatable (Rousseau, 2004). Another factor that positively affects urban birds is the increase in water supply. This is particularly evident in gardens located in arid areas where many species of desert birds can find ideal conditions for feeding and breeding. The diversity and abundance of birds is also significantly affected by vegetation characteristics. Tree cover increases the abundance of certain forest species such as woodpeckers (*Dendrocopos spp*), Eurasian Wryneck (*Jynx torquilla*), tits (*Parus spp.*) or Short-toed Treecreeper (*Certhia brachydactila*). The height of the trees does not seem to be as relevant in terms of biodiversity, although some studies consider this an important factor for the Hawfinch (*Coccothraustes coccothraustes*) and the Eurasian Golden Oriole (*Orioulus oriolus*) (Patón et al., 2012). Undoubtedly, the presence of native tree species is an important indicator to explain the abundance and diversity of birds (Fernández-Juricic, 2000b). On the other hand, in Mediterranean areas native tree species are less expensive to maintain and are more resistant to pest disease and drought conditions.

Another factor with great influence in maintaining high levels of bird diversity is shrub cover but its importance varies among different species. Some small forest birds do not use these areas much while others, such as the Common Blackbird (*Turdus merula*) strictly use the shrubs as nesting sites. Birds that are more sensitive to the presence of shrubs are basically feeding on them as the Common Nightingale (*Luscinea megarhynchus*) and typical warblers (*Sylvia spp*). For these species, the composition of the bush can be very important. Obviously native species will be the best in terms of maintaining a high bird diversity, but some foreign species may be highly suited. This is the case of many shrubs that produce winter fruits or attract insects in spring and are extensively used in gardens such as the genus *Pyracantha*, *Cotoneaster* and others. Shrubs also act as connectors between gardens so they have great importance in bird population movements. So far we have described positive factors for birds in urban areas, but obviously many factors present in the cities are markedly negative (Ditchkoff et al., 2006)). Among these factors, the highlights are chemical and noise pollution (Zannin et al., 2006). Contamination is mainly due to the high car agglomeration which provides an increase of air suspended particles, hydrocarbons, heavy metals and obviously noise. The chemical components can accumulate in the tissues of many bird species affecting its longevity and reproductive behavior (Bonier at al., 2007). Some bird species are very tolerant to noise while others are only present in the more remote areas of parks and gardens, away from traffic (Patón et al., 2012). Recent studies show that more noise sensitive species in Mediterranean areas of southern Europe, are the Goldcrest (*Regulus regulus*), the Turtle Dove (*Streptopelia turtur*), the Lesser Spotted Woodpecker (*Dendrocopos minor*), some small raptors such as the Common Buzzard (*Buteo buteo*), the Red-rumped Swallow (*Hirundo daurica*), the Common Raven (*Corvus corax*), the Golden Oriole (*Oriolus oriolus*), the Cetti's Warbler (*Cettia cetti*), the Spanish Sparrow (*Passer hispanoliensis*) and the Sardinian Warbler (*Sylvia melanocephala*) (Patón et al., 2012). Finally, we can analyze the human presence as another factor with negative impact on birds (Chávez-Zichinelli et al., 2010). Many urban birds, as sparrows, doves and starlings are widely known in their tolerance of man, but others can not adapt to human presence (Blumstein et al., 2005). In this sense, many forest species like the Common Nightingale, the European Serin (*Serinus serinus*) or the Robin (*Erithacus rubecula*) can adapt the volume, frequency or timing of the song to match the lower activity periods of people and traffic. Some species which are very abundant in rural area, are beginning to colonize the urban environment under the pressure of environmental degradation, increase in drought intensity and reduction in food supply. In Southeastern Spain, this is happening with certain corvids such as Magpies and Azure-winged Magpies (*Cyanopica cyanus*) and steppe birds as Crested Lark (*Galerida cristata*). Our present knowledge on urban areas must be complemented with more detailed studies on the environmental factors that condition the diversity and abundance of these bird communities. Today, urban areas are considered genuine anthropic ecosystems with their own internal dynamics. Paradoxically, their study allows us a better understanding of natural systems. Also, the study of urban ecosystems permits development of theoretical issues of great importance in the science of ecology. However, most studies on urban ecology have been conducted in temperate areas. There is a lack of information on tropical and Mediterranean urban areas (Suertegaray et al., 2011). In this paper, we analyze the environmental factors that determine the ecological parameters on a community level. We study 96 bird species in 35 Mediterranean sampling sites of SW Spain with varying degrees of urbanization including urban, suburban, exurban, rural, residential and wild areas.

Our results on bird species richness, evenness, diversity, dominance and abundance permit us to determine which parameters are more prevalent in urban bird communities. Consequently, we can modify these environmental factors for conservation of birds species.

2. Material and Methods

2.1. Study Areas and Sampling

The following research was conducted in a total of 35 sampling sites from SW Spain and adjacent areas of Portugal. We cover almost a complete urbanization gradient including wild, rural, exurban, suburban and urban areas (Table 1). The criteria to separate these categories of urbanization are established by Marzluff et al. (2001). These areas were selected to cover a wide range of environmental conditions for birds.

All areas were sampled at least once per month during an annual cycle. Sampling followed the recommendations of Patón et al. (2012) and correspond to fixed points recording all birds seen or heard within a circular range of 25 m around each sampling point and over 5 minutes. In the acoustic and visual identification of bird species we used the BBVA and SEO-Birdlife guide (2006). According to the size of each park we defined different sampling points separated by almost 100 m to avoid registering the same bird twice (Bart, 2004). The number of sampling points varied from 3 to 20 according to sampling area (Buckland et al., 2004). The environmental factors registered in each park are recorded in Table 2. The age of each sampling zone (AG) was determined by public information on the date of creation of the different parks and gardens.

We thought initially that this variable may be related with a successional process that attracts certain birds typical of mature forests to certain old parks. However, the majority of urban green areas are subject to an intense gardening activity that goes against this successional process. Altitude (AL), number of buildings in a circle of 500 m of radius around each sampling point (BL), tree cover (CT), distance to nearest city (DC) and park area (PA) were determined using Google-Earth. Grass (CG) and shrub (CS) covers to be partially under the tree layer were determined by both aerial photography and visually in each area.

The average level of noise (DB) of each sampling area was the annual average based on the sampling units of 5 minutes recorded simultaneously to bird sampling. We used a type II sonometer (LABBOX) following the recommendations of Malchaire (2001). The use of herbicides (HB) was measured in a semiquantitative scale of non used (0), sporadically used (1) and frequently used (2). The population of nearest city (NC) was determined using the population census of the National Institute of Statistic (INE) in Spain. Data on the population census in Portugal were obtained from the web pages of each city. The number of visitors to each sampling area (NP) was obtained recording this information during each five minute period of bird sampling. The density of old trees (OT) was recorded in each zone as the rate between the number of trees with diameter over 50 cm and signs of longevity as holes and irregular growth divided by the area of the park. The species richness of shrub (RS), trees (RT) and total woody plants (RW) was determined in each park. Finally, the number of water areas (WA) such as fountains or irrigation channels was determined.

Table 1. List of sampling sites studied in this research. These sites present very different conditions of area and altitude along an urbanization gradient Wild-Exurban-Rural-Suburban-Urban

Zone	Type	Altitude (m)	Area (ha)
Aguas Vivas (Cáceres)	Suburban	439	3.63
Bosque encantado (Avila)	Exurban	520	2.50
Campo Mayor (Portugal)	Rural	278	1.38
Campus (Badajoz)	Suburban	174	7.79
Casas Baratas (Cáceres)	Urban	452	0.28
Castañar de Casillas (Ávila)	Wild	910	23.23
Castañar del Jornillo (Avila)	Wild	950	1.87
Castelar (Badajoz)	Urban	180	1.59
Castillo de Elvas (Portugal)	Exurban	309	3.89
Chopera Cayas (Badajoz)	Rural	165	30.00
El Teso (Avila)	Exurban	580	290.11
Fresneda de las Lastras (Avila)	Rural	800	0.71
Garriga de las Lastras (Avila)	Wild	700	31.91
Infantil (Badajoz)	Urban	180	2.03
Jardines del Alcázar (Sevilla)	Urban	13	7.67
La Aliseda (Avila)	Rural	679	54.00
La Legión (Badajoz)	Urban	176	19.83
La Madrila (Cáceres)	Urban	454	6.85
Las Américas (Badajoz)	Urban	174	0.98
Las Lastras (Avila)	Exurban	700	20.15
Parque de Elvas (Portugal)	Urban	286	2.81
Parque de la Fuente del Berro (Madrid)	Urban	654	13.62
Parque de Maria Luisa (Sevilla)	Urban	16	32.45
Parque del Príncipe (Cáceres)	Urban	426	21.94
Paseo Alto (Cáceres)	Suburban	428	3.90
Paseo de Cánovas (Cáceres)	Urban	464	3.84
Paseo de Hernán Cortés (Cáceres)	Urban	443	2.83
Pinar de Sotillo de la Adrada (Avila)	Wild	1000	570.83
Pintasilgos (Portugal)	Urban	255	1.10
Plaza de Olivenza (Portugal)	Urban	262	0.91
Plaza Italia (Cáceres)	Urban	472	0.48
R66 (Cáceres)	Suburban	464	5.71
Ribera Río Cayas (Badajoz)	Wild	165	15.00
Ronda Norte (Cáceres)	Suburban	441	5.00
San Fernando (Badajoz)	Urban	177	0.81

2.2. Statistical Analysis

After the sampling of birds, the data was transformed applying rarefaction curves to avoid bias by sampling effort (Legendre and Legendre, 2003).

Table 2. Environmental parameters studied in each sampling area

Code	Description
AG	Age in years of each sampling unit determined by historic information, documents or hall records
AL	Altitude in meters determined by Google-Earth
BL	Buildings in a circle of 500 m around each sampling point
CG	Cover of grass in percentage for each sampling point determined visually
CS	Cover of shrubs in each sampling unit determined by aerial photography and visually.
CT	Cover of trees determined by aerial photography
DB	Average annual noise in decibels (dB) based on all the 5' sampling units of each park
DC	Distance to nearest city. Distance in meters to the most important and nearest city center
HB	Frequency in the use of pesticides or herbicides (0: non used; 1: sometimes; 2: frequently used)
NC	Inhabitants of nearest city
NP	Number of persons. Annual average of all sightings of people during the five minutes of sampling
OT	Density of old trees by hectare, defined as the trees with holes and a diameter over 50 cm
PA	Park area in hectares (ha)
RS	Richness of shrub species in each sampling unit
RT	Richness of tree species in each sampling unit
RW	Richness of woody species (trees and shrubs) in each sampling unit
WA	Number of water areas

The transformed data allowed the calculation of various ecological parameters of the community as the total abundance (N), the diversity of alpha-fisher (α), species richness (S), the dominance index of Berger-Parker (d), Margalef diversity (Mg), Shannon diversity (H), Shannon entropy (e^h), Simpson diversity (D) and Simpson dominance (1-D).

These parameters were used as dependent variables in a general model stepwise linear regression. Different models were tested on the significance of the coefficients (t test) in normal and homoscedasticity of waste, as the coefficient of determination (R^2) and the variation due to the model and the residual (ANOVA, F) (Mac Nally, 2003). In all the analysis the libraries BiodiversityR, vegan and stats from R statistical environment were used (R Development Core Team, 2012).

3. Results

A total of 15,744 birds from 96 species within 37 families were observed in the 35 studied areas (Table 3). The very rare woodpecker *Jynx torquilla* was detected in previous prospective studies in "Parque del Príncipe" (Cáceres, Extremadura, Spain) by D. Patón but did not appear during our sampling.

Our results show high bird richness considering that all the areas are in an urbanization gradient (MacGregor-Fors, 2010).The avian assemblages includes species under the status of vulnerable (VU), near threatened (NT) or least concern (LC) based on the Red Book of Birds of Spain (SEO, 2004).

Table 3. List of 96 species of birds detected in the sampling of 35 areas along an urbanization gradient

Scientific name	Acronym	Common name	Family	Scientific name	Acronym	Common name	
ACCIPITRIDAE	*Accipiter nisus*	ACCNIS	Eurasian Sparrowhawk	MOTACILLIDAE	*Motacilla alba*	MOTALB	White Wagtail
	Buteo buteo	BUTBUT	Common Buzzard		*Motacilla cinerea*	MOTCIN	Grey Wagtail
	Hieraaetus pennatus	HIEPEN	Booted Eagle		*Motacilla flava*	MOTFLA	Yellow Wagtail
	Milvus migrans	MILMIG	Black Kite	MUSCICAPIDAE	*Ficedula hypoleuca*	FICHYP	Pied Flycatcher
AEGITHALIDAE	*Aeghitalus caudatus*	AEGCAU	Long-tailed Tit	ORIOLIDAE	*Oriolus oriolus*	ORIORI	Golden Oriole
ALAUDIDAE	*Alauda arvensis*	ALAARV	Sky Lark	PARIDAE	*Parus ater*	PARATE	Coal Tit
	Galerida cristata	GALCRI	Crested Lark		*Parus caeruleus*	PARCAE	Blue Tit
	Lululla arborea	LULARB	Wood Lark		*Parus cristatus*	PARCRI	Crested Tit
ALCEDINIDAE	*Alcedo athis*	ALCATH	Common Kingfisher		*Parus major*	PARMAJ	Great Tit
ANATIDAE	*Anas platyrhynchos*	ANAPLA	Mallard	PASSERIDAE	*Passer domesticus*	PASDOM	House Sparrow
APODIDAE	*Apus apus*	APUAPU	Common Swift		*Passer hispanolensis*	PASHIS	Spanish Sparrow
ARDEIDAE	*Ardea cinerea*	ARDCIN	Grey Heron	PHASIANIDAE	*Alectoris rufa*	ALERUF	Red-legged Partridge
	Bubulcus ibis	BUBIBI	Cattle Egret		*Pavo cristatus*	PAVCRI	Indian Peafowl
	Egretta garzetta	EGRGAR	Little Egret	PICIDAE	*Dendrocopos major*	DENMAJ	Great Spotted Woodpecker
CERTHIIDAE	*Certhia brachydactyla*	CERBRA	Short-toed Treecreeper		*Dendrocopos minor*	DENMIN	Lesser Spotted Woodpecker
CICONIIDAE	*Ciconia ciconia*	CICCIC	White Stork		*Jynx torquilla*	JYNTOR	Wryneck
COLUMBIDAE	*Columba livia*	COLLIV	Domestic Dove		*Picus viridis*	PICVIR	Green Woodpecker
	Columba palumbus	COLPAL	Common Woodpigeon	PRUNELLIDAE	*Prunella modularis*	PRUMOD	Dunnock
	Streptopelia decaocto	STRDEC	Eurasian Collared Dove	PSITTACIDAE	*Psittacula krameri*	PSIKRA	Rose-ringed Parakeet
	Streptopelia turtur	STRTUR	European Turtle Dove	RALLIDAE	*Gallinula chloropus*	GALCHL	Moorhen
CORVIDAE	*Corvus corax*	CORCRX	Common Raven	SILVIIDAE	*Acrocephalus scirpaceus*	ACRSCI	Reed Warbler
	Corvus corone	CORCOR	Carrion Crow		*Cettia cetti*	CETCET	Cetti's Warbler
	Corvus monedula	CORMON	Eurasian Jackdaw		*Cisticola juncidis*	CISJUN	Zitting Cisticola
	Cyanopica cyanea	CYACYA	Azure-winged Magpie		*Phylloscopus collybita*	PHYCOL	Common Chiffchaff

Table 3. (Continued)

Scientific name	Acronym	Common name	Family	Scientific name	Acronym	Common name	
	Pica pica	PICPIC	Magpie		*Regulus ignicapillus*	REGIGN	Common firecrest
CUCULIDAE	*Clamator glandarius*	GARGLA	Great Spotted Cuckoo		*Regulus regulus*	REGREG	Goldcrest
	Cuculus canorus	CUCCAN	Common Cuckoo		*Sylvia atricapilla*	SYLATR	Blackcap
EMBERIZIDAE	*Miliaria calandra*	MILCAL	Corn Bunting		*Sylvia borin*	SYLBOR	Garden Warbler
	Emberiza cirlus	EMBCIR	Cirl Bunting		*Sylvia communis*	SYLCOM	Common Whitethroat
ESTRILDIDAE	*Taeniopygia guttata*	TAEGUT	Zebra Finch		*Sylvia hortensis*	SYLHOR	Western Orphean Warbler
FALCONIDAE	*Falco naumanni*	FALNAU	Lesser Kestrel		*Sylvia melanocephala*	SYLMEL	Sardinian Warbler
	Falco tinunculus	FALTIN	Common Kestrel	SITTIDAE	*Sitta europaea*	SITEUR	European Nuthatch
FRINGILLIDAE	*Carduelis cannabina*	CARCAN	Linnet	STRIGIDAE	*Asio flammeus*	ASIFLA	Short-eared Owl
	Carduelis carduelis	CARCAR	Goldfinch		*Athene noctua*	ATHNOC	Little Owl
	Carduelis chloris	CARCHL	Greenfinch		*Strix aluco*	STRALU	Tawny owl
	Carduelis spinus	CARSPI	Siskin	STURNIDAE	*Sturnus unicolor*	STUUNI	Spotless Starling
	Coccothraustes coccothraustes	COCCOC	Hawfinch	TROGLODYTIDAE	*Troglodytes troglodytes*	TROTRO	Eurasian Wren
	Fringilla coelebs	FRICOE	Common Chaffinch	TURDIDAE	*Erithacus rubecula*	ERIRUB	Robin
	Serinus serinus	SERSER	European Serin		*Luscinia megarhynchos*	LUSMEG	Rufous Nightingale
HIRUNDINIDAE	*Delichon urbicum*	DELURB	House Martin		*Phoenicurus ochruros*	PHOOCH	Black Redstart
	Hirundo daurica	HIRDAU	Red-rumped Swallow		*Phoenicurus phoenicurus*	PHOPHO	Common Redstart
	Hirundo rustica	HIRRUS	Barn Swallow		*Saxicola rubetra*	SAXRUB	Whinchat
LANIIDAE	*Lanius excubitor*	LANEXC	Great Grey Shrike		*Saxicola torquatus*	SAXTOR	Common Stonechat
	Lanius senator	LANSEN	Woodchat Shrike		*Turdus iliacus*	TURILI	Redwing
LARIDAE	*Larus ridibundus*	LARRID	Black-headed Gull		*Turdus merula*	TURMER	Blackbird
MEROPIDAE	*Merops apiaster*	MERAPI	European Bee-eater		*Turdus philomelos*	TURPHI	Song Thrush
MOTACILLIDAE	*Anthus pratensis*	ANTPRA	Meadow Pipit		*Turdus viscivorus*	TURVIS	Mistle Thrush
	Anthus trivialis	ANTTRI	Tree Pipit	UPUPIDAE	*Upupa epops*	UPUEPO	Hoopoe

Under the VU category are the species *Falco naumanni* and *Streptopelia turtur* that represent 0.51% and 0.08% of the 14,744 birds sampled. As NT are *Alcedo athis* (0.06%), *Asio flammeus* (0.06%), *Lanius senator* (0.01%) and *Milvus migrans* (0.06%). Under LC status only *Sylvia hortensis* (0.03%) appears. In consequence, the most rare birds belong to the NT group that represents the most severe status of protection.

In our sampling, exotic species such as Zebra finch (*Taeniopygia guttata*) with a 0.01% of abundance and Rose-ringed Parakeet (*Psittacula krameri*) with a 0.05% were detected. The ten most abundant species were the House Sparrow (19.43%), the Blackbird (8.18%), the Spotless Starling (6.65%), the Robin (5.41%), the Serin (4.93%), the Domestic Dove (4.91%), the Eurasian Collared Dove (4.75%), the Goldfinch (4.72%), the Azure-winged Magpie (3.55%) and the House Martin (2.64%).

These species represent 65.17% of the avian community indicating the great dominance that is typical of urban areas. The differences in environmental factors along the gradient of urbanization are shown in Table 4.

Obviously, the most unique factor between zones in the gradient is the number of buildings within a radius of 500 m around each point (BL), followed by the distance to town (DC), noise (DB), number of persons in the sampling (NP), number of people in the nearest city (NC) and various vegetation factors (OT, CS, CT).

Table 4. Environmental factors described in Table 2 along an urbanization gradient Wild-Exurban-Rural-Suburban-Urban. The test of Kruskal-Wallis show the significance of these factors in each group of areas

Exurban	Rural	Suburban	Urban	Wild	KW	
AG	254.3	97.5	20.0	128.6	103.0	3.73 ns
AL	527.3	480.5	389.2	299.1	745.0	8.68 ns
BL	71.3	45.0	255.2	360.8	16.8	24.23***
CG	23.8	20.0	1.2	1.9	5.2	5.10 ns
CS	4.0	0.3	0.4	2.1	22.0	13.11*
CT	34.1	35.1	1.5	4.0	71.0	16.00**
DB	49.5	41.4	53.7	58.4	37.4	17.98**
DC	311.0	570.9	0.0	0.0	935.9	23.98***
HB	0.0	0.3	0.2	0.1	0.0	2.19 ns
NC	10949.5	42835.8	105943.2	354633.5	33634.8	14.88**
NP	0.2	0.2	0.4	1.5	0.1	16.43**
OT	0.8	0.8	0.0	0.5	1.0	11.55*
PA	79.2	21.5	5.2	7.1	128.6	6.24 ns
RS	10.8	4.3	6.0	10.0	4.6	3.29 ns
RT	17.5	5.3	8.6	20.2	4.8	7.35 ns
RW	28.3	9.5	14.6	30.2	9.4	5.84 ns
WA	4.0	2.3	0.8	2.5	2.4	2.39 ns

Ns: non-significant differences. *: p-value < 0.05; **: p-value < 0.01; ***: p-value < 0.001.

With respect to the differences in ecological parameters of bird communities along gradient we show in Table 5 that the significant parameters were total bird abundance (N), the Margalef diversity (Mg) and the alpha-fisher diversity (α).

Consequently we determine the multiple regression models only for the environmental factors and ecological parameters that are significant along the urbanization gradient. For parameter N we obtain the following equation:

$$N = 0.379 * DB$$

This model shows a R^2 of 0.83, with non-normality of residuals (W=0.92; p-value=0.02) and heteroscedasticity of residuals (BP=0.10; p-value = 2.2e-16). The parameter Mg responds in a more complex manner to environmental factors along the urbanization gradient with this model:

$$Mg = -0.006*BL + 0.09*DB + 0.001*DC$$

The equation presents an R^2 of 0.94, with residuals normality (W = 0.99; p-value = 0.92) and homoscedasticity (BP = 2.08; p-value = 0.35). The model for α is the following:

$$\alpha = 0.04*DB + 0.004*DC + 2.49*OT$$

This equation presents an R^2 of 0.82, with non-normality of residuals (W = 0.92; p-value = 0.02) and homoscedasticity (BP = 4.17; p-value = 0.12). Although the stepwise regression models are mainly used to determine which factors affect each ecological parameter, we can use the model to calculate the Margalef diversity for predictive purposes because it has residual normality and homoscedasticity.

Table 5. Differences in ecological parameters total bird abundance (N), alpha-fisher diversity (α), species richness (S), Berger-Parker dominance (d) , Margalef diversity (Mg), Shannon diversity (H), Shannon entropy (e^h), Simpson diversity (D) and Simpson dominance (1-D). The test of Kruskal-Wallis show the significance of these factors in each group of areas

Ecological parameter	Exurban	Rural	Suburban	Urban	Wild	KW
N	10.5	15.3	24.2	23.7	7.9	14.72 **
d	0.4	0.5	0.4	0.5	0.4	9.36 ns
S	5.1	5.3	6.2	4.5	4.6	3.81 ns
Mg	4.1	3.9	3.9	2.7	4.1	16.18 **
α	6.3	5.5	4.4	3.0	7.4	15.72 **
H	1.4	1.3	1.5	1.1	1.3	6.65 ns
e^h	4.2	3.8	4.6	3.2	3.8	6.47 ns
D	0.7	0.7	0.7	0.6	0.7	6.78 ns
1-D	4.0	3.3	4.0	2.9	3.8	8.59 ns

Ns: non-significant differences. **: p-value < 0.01.

Noise is the only factor affecting the total abundance (N), alpha-fisher diversity (α) and Margalef diversity (Mg). The distance to the city is a factor with a very important effect on diversity but not on abundance. The presence of old trees is also a relevant variable in terms of bird diversity. The models leave out other aspects such as coverage of tree or shrub vegetation, pollution levels or even the presence of people.

4. DISCUSSION

The results clearly show that the ecological parameters of the bird community which respond significantly to gradients of urbanization are limited. Only the total number of birds (N) and alpha-fisher (α) and Margalef (Mg) diversities are appropriate to show changes between wild, exurban, rural, suburban and urban areas. Without a doubt, the bird communities present differences in other ecological indexes but its high variance could account for this lack of significance. Many ecological indexes are from coarse-grained and therefore inappropriate to detect gradual changes in gradients of urbanization. The classification proposed by Marluff et al. (2001) seems highly functional and easy to apply in the context of the urban ecology of birds. Our results indicate that vegetation aspects are not as important as we anticipated. However other studies suggest otherwise (Carbó-Ramirez and Zuria, 2011; Daniels and Kirkpatrick, 2006), indicating that we should expand the sampling to include more parks, especially young ones with less tree cover. However, our study emphasizes that the most important limiting factor for urban birds in terms of abundance and diversity is urban noise. Moreover, noise acts as a real and invisible barrier for many bird species. A previous study has shown that under 50 dB many species can colonize the urban environments (Patón et al., 2012). This threshold of 50 dB separates areas with high species richness from the rest. However bird richness might not be significantly different between areas due to the large variability of this parameter. In spite of using a stepwise linear regression restricted to significant independent (factors) and dependent (ecological parameters) variables we believe that sampling should be increased to more areas for more accurate conclusions. In fact, there is high variability between urban areas and any study aimed at obtaining useful mathematical models should be extended to the maximum variability of environmental conditions. Really we believe that the interference between environmental factors can be very marked in these urban environments. For example, a high level of tree cover should be positive as long as noise does not affect bird abundance or richness. It is important that we have perfectly clear the interference and hierarchy of importance of all the environmental factors that affect urban birds. Then we can design the ideal park for birds with conservation criteria. No doubt other benefits will become associated with this design of urban parks such as the use of these areas in environmental education programs. The list of species studied in the present research is large for Mediterranean areas and includes even endangered species. Consequently, urban areas have the potential to be considered small reserves for birds. Although much additional research must be conducted, we can estimate, based on this and other related studies, that the most biodiverse urban parks would be those with a low level of predators, high shrub cover, high presence of old trees, large number of native plant species, low noise level and which are located in the periphery of medium-sized towns (Fang and Ling, 2005; Ortega-Alvárez and McGregor Fors, 2009; Peris and Pescador, 2004). The southwest of the Iberian Peninsula has many small towns that could host greenbelts suitable for urban birds. These can be transitional areas between the urban and the natural environment that act as ecological corridors (Sandström et al., 2006). These parks can be multipurpose areas for chemical and acoustic decontamination, places of environmental education and CO_2 sinks. We believe it is necessary to open a debate between politicians, managers and scientists to enable a better urban design incorporating these ideas.

ACKNOWLEDGMENTS

We want to thank the Association of Friends of the "Parque del Príncipe" (Cáceres) and the Parks Services of municipalities of Cáceres, Badajoz, Madrid and Sevilla for the facilities extended for this study. Several courses of students of Environmental Sciences worked selflessly in some of the censuses of the study. We thank the scientists cited in the references for their useful contributions to this area of research. We also thank J. Elliott Casal for his English corrections on this manuscript. Part of this study was financed with funds for practical classes at the University of Extremadura (SW Spain).

REFERENCES

Akinnifesi, F. K., Sileshi, G. W., Ajayi, O. C., Akinnifesi, A. I., De Moura, E. G., Linhares, J. F. P., and Rodrigues, I. (2010). Biodiversity of the urban homegardens of São Luís city, Northeastern Brazil. *Urban Ecosystems* 13, 129-146.

Aurora, A. L., Simpson, T. R., Small, M. F., and Berder, K. C. (2009). Toward increasing avian diversity: urban wildscapes programs. *Urban Ecosystems* 12, 347-358.

Bart, J., Fligner, M. A. and Notz, W. I. (2004). *Sampling and statistical methods for behavioral ecologists*. Cambridge University Press, United Kingdom, 330 pp.

BBVA and SEO-Birdlife (2006). Descubre las aves por sus cantos (Discover the birds by their songs). Fundación BBVA and SEO-Birdlife, Madrid, Spain [in Spanish].

Blumstein, D. T., Fernández-Juricic, E., Zollner, P. A., and Garity, S. C. (2005). Inter-specific variation in avian responses to human disturbance. *Journal of Applied Ecology* 42, 943-953.

Bonier, F., Martin, P. R. and Wingfield, J. C. (2007). Urban birds have broader environmental tolerance. *Biological Letters* 3, 670-673.

Buckland, S. T., Anderson, D. R., Burnham, K. P., Laake, J. L., Borchers, D. L., and Thomas, L. (2004). *Advanced distance sampling*, Oxford University Press, US, 997 pp.

Carbó-Ramírez, P. and Zuria, I. (2011). The value of small urban greenspaces for birds in a Mexican city. *Landscape and Urban Planning* 100, 213-222.

Caula, S. A., Sirami, C., Marty, P., and Martin, J. L. (2010). Value of an urban habitat for the native Mediterranean avifauna. *Urban Ecosystems* 13, 73-89.

Chávez-Zichinelli, C. A., MacGregor-Fors, I., Talamás, P., Valdéz, R., Romano, M. C., and Schondube, J. E. (2010). Stress responses of the House Sparrow (*Passer domesticus*) to different urban land uses. *Landscape and Urban Planning* 98, 183-189.

Daniels, G. D. and Kirkpatrick, J. B. (2006). Does variation in garden characteristics influence the conservation of birds in suburbia?. *Biological Conservation* 133, 326-335.

Ditchkoff, S. S., Saalfeld, S. T. and Gibson, C. J. (2006). Animal behaviour in urban ecosystems: Modifications due to human-induced stress. *Urban Ecosystems* 9, 5-12.

Fang, C. F. and Ling, D. L. (2005). Guidance for noise reduction provided by trees belts. *Landscape and Urban Planning* 71, 29-34.

Fernández-Juricic, E. (2000a). Bird community composition patterns in urban parks of Madrid: The role of age, size and isolation. *Ecological Research* 15, 373-383.

Fernández-Juricic, E. (2000b). Avifaunal use of wooded streets in an urban landscape. *Conservation Biology* 14(2), 513-521.

Fernández-Juricic, E. and Jokimäki, J. (2001). A habitat island approach to conserving birds in urban landscapes: case studies from southern and northern Europe. *Biodiversity and Conservation,* 10, 2023-2043.

Legendre, P., Legendre, L. (2003). *Numerical ecology*, Elsevier Science, The Netherlands, 853 pp.

MacGregor-Fors, I. (2010). How measure the urban-wildland ecotone: redefining peri-urban areas. *Ecological Research* 25, 883-887.

Mac Nally, R. (2000). Regression and model building in conservation biology, biogeography and ecology: the distinction between and reconciliation of 'predictive' and 'explanatory' models. *Biodiversity and Conservation* 9, 655-671.

Malchaire, J. (2001). Sound measuring instruments. In: Goelzer, B., Hansen, C., Sehrndt, G. (Eds.), *Ocuppational exposure to noise: Evaluation, prevention and control.* World Health Organisation, Federal Institute for Occupational Safety and Health, Dortmund, pp. 125-140.

Marzluff, J. M., Bowman, R. and Donnelly, R. (2001). A historical perspective on urban bird research: trends, terms, and approaches. In: *Avian ecology and conservation in an urbanizing world.* 1-17 pp. Ed. Kluwer Academic Press.

Ortega-Álvarez, R. and McGregor-Fors, I. (2009). Living in the big city: Effects of urban land-use on bird community structure, diversity and composition. *Landscape and Urban Planning* 90, 189-195.

Patón, D., Romero, F., Cuenca, J., and Escudero, J. C. (2012). Tolerance to noise in 91 bird species from 27 urban gardens of Iberian Peninsula. *Landscape and Urban Planning* 104, 1-8.

Peris, S. J. and Pescador, M. (2004). Effects of traffic noise on paserine populations in Mediterrenean wooded pastures. *Applied Acoustic* 65, 357-366.

R Development Core Team (2012). R Installation and administration. GNU Free Software Foundation. US, 25 pp.

Rousseau, J. (2004). Avian diversity, assemblages and use of vegetation, mainly by shrub-nesters, in an urban ecosystem. *Ph D Thesis*. Mc Gill University, Quebec, Canada, 272 pp.

Sandström, U. G., Angelstam, P., Mikusinski, G. (2006). Ecological diversity of birds in relation to the structure of urban green space. *Landscape and Urban Planning* 77, 39-53

SEO (2004). *Libro rojo de las aves de España* (Red data book of birds of Spain). Eds: Madroño, A., González, C., Atienza, J. C. Sociedad Española de Ornitología, Madrid, Spain [in Spanish], 452 pp.

Shochat, E., Lerman, S. B., Anderies, J. M., Warren, P. S., Faeth, S. H., and Nilon, C. H. (2010). Invasion, Competition, and Biodiversity Loss in Urban Ecosystems. *BioScience.* 60(3), 199-208

Suertegaray, C., Burger, M. I. and Magnusson, W. E. (2011). Bird diversity in a subtropical South-American City: effects of noise levels, arborisation and human population density. *Urban Ecosystem* 14(3), 341-360.

United Nations Population Division (2009). World urbanization prospects. U.N. Department of Economical and Social Affairs, New York, US, 7 pp.

Zannin, P. H. T., Coelho Ferreira, A. M., Szeremetta, B. (2006). Evaluation of noise pollution in urban parks. *Environmental Monitoring Assessment* 118, 423-433.

In: Birds
Editors: Lucas Ruiz and Franco Iglesias

ISBN: 978-1-62948-104-3

Chapter 9

The Impact of Landscape Configuration and Competitors on the Hooded Vulture *Necrosyrtes monarchus* Temminck 1823 in Southern Ghana

Michael O'Neal Campbell[#]
Department of Geography, Thompson Rivers University, Kamloops, BC, Canada

Abstract

The Hooded Vulture, the smallest and commonest vulture in West Africa, is also possibly the most adaptive African vulture, evidenced by its tolerance of human proximity and high urban presence. However, very few studies assess its role as a dominant urban and rural scavenger in West African contexts outside the range of the larger vultures, rather focusing on its relationship with the larger vultures in the Eastern and Southern African plains. This chapter examines the relationships between Hooded Vultures, competing avian scavengers (black kites and pied crows) and rural/urban landscape configuration in the little researched southern half of Ghana, West Africa. Similar to crows and kites, hooded vulture numbers were positively correlated with settlement populations, and these birds had a stronger presence in urban than rural contexts. Hooded vultures had higher correlations with organic waste than crows or kites, and were the commonest avian scavengers in urban areas with high organic waste (abattoirs, markets), but less common than pied crows in residential and business districts with low organic waste. Vultures, kites and crows were commoner in urban areas near dense forest (including dense, urban green spaces), than in urban areas near lighter vegetation. Vultures nesting was correlated with dense rural forests, and to a lesser extent with dense, urban green spaces. Vultures were active in the late evening, early dawn and night in urban areas. In rural areas, vulture numbers were less correlated with vegetation density than in urban areas, and town proximity was positively correlated with vulture

[#] Corresponding author: Michael O'Neal Campbell. Department of Geography, Thompson Rivers University, 900 McGill Road, Kamloops, BC, Canada. E-mail: ecol55@hotmail.com.

numbers. Conflicts between vultures, kites, crows and feral dogs were common, and no species was significantly dominant. The hooded vulture therefore has some specialist tendencies, despite its common portrayal as a successful generalist.

Introduction

The hooded vulture *Necrosyrtes monarchus* Temminck 1823, a large, diurnal African raptor, is the commonest and smallest West African vulture, present in both forest and savanna, and urban areas (Brown et al. 1982; Cramp and Pernis 1977 - 94; Sibley et al. 1990; Dowsett and Forbes-Watson 1993; Houston 2001). It is described as 67-70 cm long, a small, scruffy-looking, mostly brown vulture, with long thin bill, bare crown, face and foreneck, conspicuous ear-holes, and downy nape and hind neck, that perches hunched with wings drooping (Birdlife International 2013a). Although it is very common in West Africa, it is less studied there than in East and South Africa (Newton 1979; Mundy et al. 1997; Lees and Christie 2001; Virani et al. 2011). In particular, there are few detailed studies of hooded vulture ecology and distribution in the West African country of Ghana (8 00 N, 2 00 W; area 239,460 sq km; human population 20,757,032 (CIA 2004) Figure 1). Ghana, a country "virtually unknown to bird watchers", lacks studies of even very common species such as the hooded vulture (Finn (2004). This species is the only common vulture in the largely forested and urbanised southern half of the country, other larger vultures restricted to the savanna dominated northern half (Bannerman 1930; Campbell 2009). These other species are the white backed vulture *Gyps africanus* Salvadori 1865, white headed vulture *Trigonoceps occipitalis* Burchell 1824 and Egyptian vulture *Neophron percnopterus* Linnaeus 1758 (Mundy et al. 1997; Thiollay 2006; Birdlife International 2013a,b,c,d). Current studies of the hooded vulture neglect its status as possibly the dominant avian scavenger in southern Ghana (Campbell 2009). Most studies briefly discuss mundane topics such as its small size, low competitive status, tolerance for humans (all relative to other vultures) and traditional cultural viewpoints (Newton 1979; Ntimoah Baidu 1987; Ferguson-Lees and Christie 2001; Houston 2001; Butchart and Roche 2003; Iddi 2004). Few detailed, recent studies have investigated the impacts of landcover, resource activities (abattoirs, rubbish dumps, markets, central business districts and residential areas in urban areas, and rural vegetation, fires, transport networks and farming in rural areas) and other scavengers on vulture ecology. Due to the absence of other vulture species from southern Ghana, the competing scavengers are usually pied crows *Corvus albus* Muller 1776, black kites *Milvus migrans* Boddaer 1783 and feral dogs *Canis lupus familiaris* Linnaeus 1758 (Campbell 2009). The objective of this study is to test the hypothesis that the landcover pattern of the southern Ghanaian settlements affects hooded vulture presence, this varying from presence in rural areas.

Methodology

The Point Count Method

The survey comprised a point count method, as described by Tewksbury (1998) and Bayne and Hobson (2000).

One hundred and five observation points were taken, each visited for at least 15 minutes at least 10 times, during the period September 1985 - September 1993, May 1995- September 1996, and April – June 2000.

All sightings of hooded vultures, pied crows and black kites were recorded. Where feeding and foraging occurred, inter- and intra - species interactions, and those with feral dogs were also monitored.

The statistical analysis used regression analysis to test the impact of landscape configuration and associated animal species on the behaviour of hooded vultures. For this analysis, the dependent variable was vulture numbers and the independent variables were numbers of kites and crows and components of the landcover. Kite and crow numbers were also regressed as dependent variables against independent landcover variables.

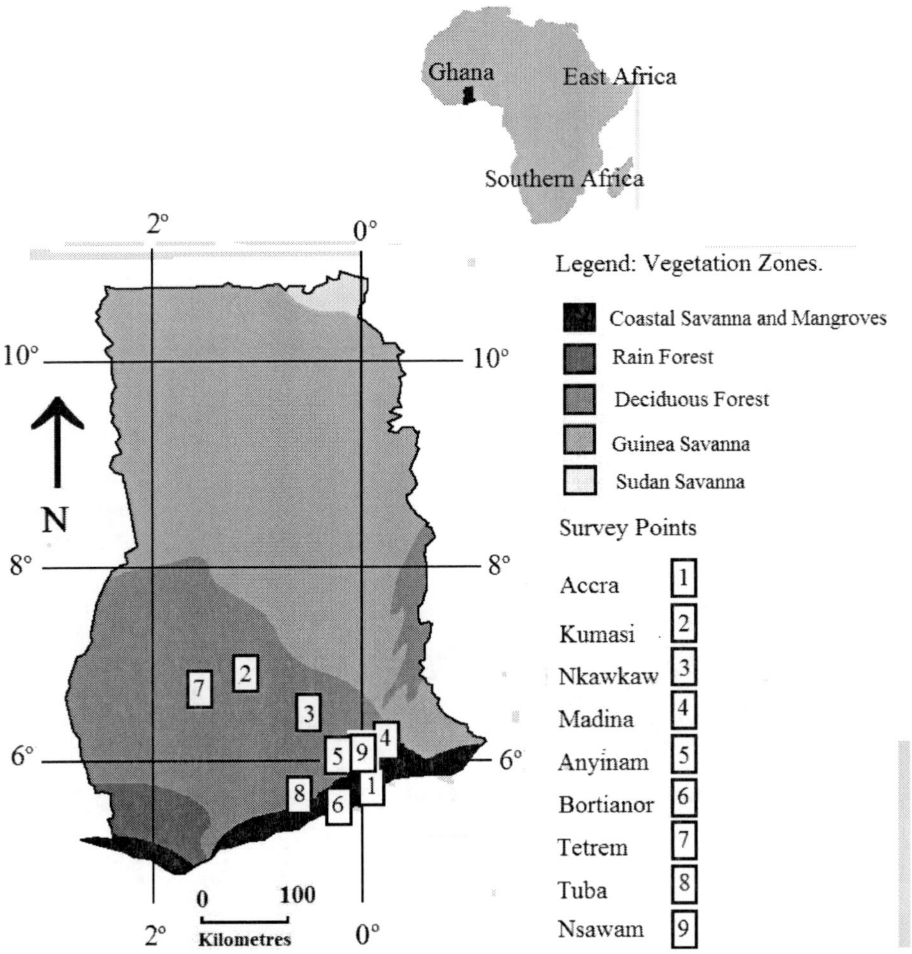

Figure 1. The survey points in the vegetation zones of Ghana. Note the location of Ghana in the map of Africa.

Urban Locations

Urban areas were defined as settlements with populations above 5,000, with dominant secondary and tertiary activities; other settlements were defined as villages (Dickson and Benneh 1995). Human population figures were derived from the Ghana Census (2000). Larger settlements were selected deliberately, as these represented the largest concentrations of human population in southern Ghana. The choice of the smaller settlements was random, taken as representative of both small nearby satellite settlements and distant towns more isolated in the rural landscape (Table 1, Figure 1). Urban survey points were chosen randomly within the proximity of slaughterhouses or abattoirs, rubbish dumps, markets, business districts, residential areas and green spaces. Abattoirs varied in size from factory- sized buildings in the large cities (Accra and Kumasi) capable of killing herds of cattle daily, to smaller butcher's shops in all settlements, where mostly small livestock (sheep, goats, pigs, chickens) were slaughtered. Rubbish dumps in larger cities comprised both large open areas (up to eight hectares in area) areas for landfills and smaller neighbourhood truck trailer type dumps. Most dumps in smaller settlements tended to have more organic waste, due to the less effective sanitation restrictions. Markets comprised open air socio-economic complexes of varying sizes (from ten hectares to less than ten square metres), most producing organic waste. Other surveyed areas with occasional organic waste were business districts, residential areas, green spaces, potholed streets and gutters.

Rural Locations

Eighty random locations were also taken in the semi-deciduous forest and 30 in the coastal savnana. These were classified according to: (1) forest; (2) savanna (mixed forest, shrub and grass mosaics); and (3) settlement proximity. The dominant vegetation within one square kilometre of the survey point was used for classification. Roads in these areas were also monitored for road kills; as were areas of bush clearance, forest fires and termite flights. These have been described as bird attractants (Ferguson–Lees and Christie 2001). The dominant forest vegetation was the deciduous forest of central Ghana (Figure 1). The structure of these stands were approximately 25 percent contiguous canopy 35 to 50 metres high, and fragmented tree stands mixed with saplings and farms.

Table 1. Characteristics of Settlements in the Survey

Settlement	Coordinates	Population	Nearby Vegetation	Points
Accra	5.56°N 0.20°W	1,755,000	Savanna	35
Kumasi	6.69°N; 1.63°W	785,000	Forest	35
Nkawkaw	6.55°N; 0.78°W)	55,545	Forest	5
Madina	5.68°N; 0.17°W	47,101	Savanna	5
Nsawam	5.81°N; 0.36°W	33,876	Savanna	5
Anyinim	6.37°N; 0.55°W	8,353	Forest	5
Bortianor	5.51°N; 1.63°W	3,987	Savanna	5
Tetrem	6.55 °N; 1.92°W	3,497	Forest	5
Tuba	5.52°N;0. 34°W	695	Savanna	5

Dominant species were *Triplochiton scleroxylon* K.Schum. 1900, *Terminalia ivorensis* A.Chev 1909, *Heriteria utilis* Sprague 1909, *Entandrophragma cylindricum* Sprague 1910, *Samarea saman* 1916, *Ceiba pentandra* (L.) Gaertn. 1791 and *Piptadeniastrum africanum* Hook.f 1955. The savanna was the coastal savanna, which was a mixture of fragmented stands of deciduous forest trees, mixed with neem *Azadirachta indica* A.Juss 1890 and *Mangifera indica* L. 1753, *Cassia siamea* Lam. 1785, *Ficus thonningii* Blume 1836, *Millettia thonningii* (Schum. and Thonn.) Baker 1948, *Antiaris toxicaria* Lesch.1811, *Fagara zanthoxyloides* Lam. 1758 and grasses (dominated by *Andropogon gayanus* L.1758, *Panicum maximum* Jacq. 1781, and *Sporobulus* species).

Results

Hooded vultures were more common in urban contexts (on average 4.8 birds per point) than in forested (2.1) or savanna areas (1.6). Black Kites (urban = 0.6, forest = 0.7, savanna = 0.3 per point) and pied crows (urban = 4.2, forest = 2.8, savanna =1.4) were also slightly commoner in urban areas than forest and savanna. In terms of absolute numbers, in urban areas, there was a weakly positive relation between vulture numbers and the population of the settlements ($r = 0.035$, $r^2 = 0.001$, $F = 3.130$, $p > 0.05$), which contrasted with the weakly negative relation between black kite numbers and town population ($r = -0.063$, $r^2 = 0.004$, $F = 0.425$, $p > 0.05$), and that of pied crows ($r = -0.168$ $r^2 = 0.028$, $F = 3.139$, $p > 0.05$). The explanation of variance of hooded vulture population density in urban settings is shown in Figure 2.

All three bird species in urban areas showed positive correlations with the dense vegetation (forest or dense forest-savanna mosaics), as opposed to open, shrub-grass dominated savanna (hooded vulture $r = 0.373$ $r^2 = 0.139$, $F = 17.470$, $p < 0.05$; black kite 0.301 $r^2 = 0.091$, $F = 10.756$, $p < 0.05$; pied crow 0.129, $r^2 = 0.017$, $F = 1.820$, $p < 0.05$). The pied crow showed the greatest balance between open areas and forested vegetation. Hooded vulture numbers had the highest positive correlation with the size of nearby inner city green spaces (hooded vultures 0.261, $r^2 = 0.064$, $F = 7.900$, $p < 0.05$; black kite 0.085, $r^2 = 0.007$, $F = 0.794$, $p > 0.05$; pied crow 0.139, $r^2 = 0.019$, $F = 2.124$, $p > 0.05$).

The number of vultures perching were also significantly, but weakly correlated with the size of nearby green spaces ($r = 0.252$, $r^2 = 0.064$, $F = 7.346$, $p < 0.05$) and the density of nearby vegetation (contiguous or fragmented forest) in peri-urban areas ($r = 0.268$, $r^2 = 0.072$, $F = 8.376$, $p < 0.05$), but only weakly negatively correlated to nesting areas ($r = -0.054$, $r^2 = 0.003$, F 0.313, $p > 0.05$). The numbers of vultures perching for food was significantly correlated with vulture numbers in flight ($r = 0.851$, $r^2 = 0.724$, $F = 284.237$, $p < 0.05$) and also with the presence of black kites ($r = 0.308$, $r^2 = 0.094$, 11.341, $p < 0.05$) and pied crows ($r = 0.358$, $r^2 = 0.128$, 15.900, $p < 0.05$). Passing automobiles and people, more frequent in larger towns, generally disturbed vultures and crows within about two to three metres, and kites at about 3.5 to four metres.

The strongest explanations for variance in bird numbers in urban areas were landuse variation: abattoirs, rubbish dumps, markets, business districts and residential areas. At the 25 abattoirs monitored, hooded vultures were the most common species (on average 11.1 birds per point, 5.4 perching), compared with black kites (0.9) and pied crows (4.7).

At the twenty-three rubbish dumps surveyed, hooded vultures were again the commonest species (4.1 with 1.5 perching, black kites 0.7 and pied crows 3.2). At the twenty-one markets surveyed, there were an average of 5.5 vultures (1.3 perching), 0.4 kites and 3.2 crows. Fewer birds were recorded in the twenty-two business districts (1.0 vultures, none perching, 0.3 kites, 1.9 crows) and the nineteen residential areas (0.7 vultures with 0.1 perching, 0.4 kites, 2.4 crows). In the built up areas, crows exhibited greater agility in accessing scraps from more precarious areas. These included windowsills, balconies, gardens, roads and sidewalks, and rubbish dumps during dumping. They also appeared to have a wider diet; fruits and vegetation, as well as the meat favoured by other scavengers.

Ranking the urban areas according to organic matter available (abattoirs = 5, rubbish dumps = 4, markets = 3, business districts = 2, residential areas = 1) and regressing this ranking with bird numbers produced the highest figures for vulture presence ($r = 0.560$, $r^2 = 0.313$, $F = 49.245$, $p < 0.05$) and vulture perching ($r = 0.481$, $r^2 = 0.231$, $F = 32.444$, $p < 0.05$), with lower correlations with pied crows ($r = 0.376$, $r^2 = 0.141$, $F = 17.813$, $p < 0.05$) and black kites ($r = 0.251$, $r^2 = 0.063$, $F = 7.254$, $p < 0.05$). Where the three species were sharing feeding sites, approximately twenty percent of the encounters between hooded vultures and pied crows resulted in conflicts (with crows mobbing and chasing vultures) as did six percent of the encounters between vultures and black kites. Of the 13 interactions recorded between vultures and feral dogs in urban settings, six resulted in conflicts, with even numbers favouring the two species. These resulted in winning scraps, but not leaving behaviour.

Of the 73 nests recorded, there were only a few records of vultures nesting in urban areas, all of these being in green spaces (urban green spaces 11 percent, nearby peri-urban or rural forests/woodlots 18 percent, and distant (10 km +) forests/savanna woodstands 71 percent). Nesting behaviour was moderately correlated with green space area ($r = 0.219$, $r^2 = 0.048$, $F = 5.416$, $p < 0.05$) and the density of the treed growths ($r = 0.218$, $r^2 = 0.048$, $F = 5.378$, $p < 0.05$). Sixty eight percent of the non-nesting vultures seen roosting at dusk or dawn were in nearby forests/wooded savanna, 29 percent in distant forests, and three percent were in urban built up areas (in single trees and roof tops).

All the roosting black kites were in distant forests. Of the 60 pied crow nests observed, the majority were in nearby forest savanna (72 percent); the rest were in distant forest (12 percent) and urban areas (16 percent). Where pied crows roosted close to vultures, conflicts were observed. About 23 percent of these resulted in the expulsion of vultures from roosts. In urban roosting areas and well-lit markets and dumps, some vultures and crows remained active in low flight and feeding after dark (7 pm +). There were vultures in flight at dawn from rural roosting areas to urban sites, usually ten to twenty minutes before or after sunrise, around 6 am at low altitudes of five to 15 metres. About 46 percent of these were in the direction of known abattoirs, rubbish dumps and markets. Similar behaviour was observed in crows (about 30 percent), although this was less directional, and occasionally in kites (about six percent of the sightings).

In the rural areas, there was a moderate positive correlation between vegetation density and bird numbers, although the figure for vultures was not significant compared to urban areas near to forest (vultures $r = 0.105$, $r^2 = 0.011$, $F = 1.219$, $p > 0.05$; kites $r = 0.225$, $r^2 = 0.051$, $F = 5.823$, $p < 0.05$; crows $r = 0.218$, $r^2 = 0.0.048$, $F = 5.448$, $p < 0.05$). Numbers of all species increased in rural areas near to towns, but for kites this was not significant (vultures $r = 0.377$, $r^2 = 0.142$, $F = 18.059$, $p < 0.05$; kites $r = 0.181$, $r^2 = 0.033$, $F = 3.671$, $p > 0.05$; crows $r = 0.264$, $r^2 = 0.070$, $F = 8.148$, $p < 0.05$).

This was positively correlated with the size of the nearby towns (vultures r = 0.407, r^2 = 0.166, F = 21.686, $p < 0.05$; kites r = 0.200, r^2 = 0.040, F = 4.539, $p < 0.05$; crows r = 0.471, r^2 = 0.045, F = 5.066, $p < 0.05$). Town proximity and size also positively affected perching and feeding behaviour in rural areas (r = 0.213, r^2 = 0.045, F = 5.195, $p < 0.05$; and r = 0.247, r^2 = 0.061, F = 7.076, $p < 0.05$ respectively) as larger towns frequently had waste areas and poorer sanitation in peri-urban areas.

In rural areas, vulture perching was more strongly correlated with the presence of other species (black kite r = 0.471, r^2 = 0.222, F = 31.038, $p < 0.05$; pied crow r = 0.489, r^2 = 0.239, F = 34.195, $p < 0.05$), than in the urban areas. Also, in rural areas, 26 percent of the encounters between hooded vultures and pied crows resulted in conflict, as did eleven percent of the encounters between vultures and black kites. Of the eight interactions recorded between feral dogs and vultures in rural settings five resulted in conflicts, with the vultures prevailing in two and the dogs in the rest, with one resulting in the vultures leaving altogether.

Bushfires, termite flights (mostly *Macrotermes* species), bush clearance and occasional road kills all attracted larger numbers of crows than vultures or kites. In bushfire areas (started by people for land clearance) the figures were greatest for black kites (3.1 per fire, pied crows 1.8, hooded vultures 1.7), as were termite flights (2.6, pied crows 1.2, vultures 0.2). For bush clearance (with machetes and/or tractors) pied crows had the highest presence (5.3, kites 1.5, vultures 0.4). The eight road kills recorded in forest (four maxwell's duikers *Cephalophus maxwellii* H.Smith.1827, a black duiker *Cephalophus niger* Gray 1846, two cane rats *Thryonomys swinderianus* Temminck 1827 and a palm civet *Nandina binotata* Gray 1830) were attended by 16 pied crows, eight hooded vultures and two black kites. There were two conflicts between the vultures and crows, one an airborne pursuit of a vulture and another brief ground fight, where the vulture prevailed. In savanna areas the figures were similar, but pied crows were commoner at fires (7.3, kites 0.7, vultures 0.4), termite flights (5.2, kites 0.2, vultures 0.2) and bush clearance (4.1, kites 0.1, vultures 0.1). The six road kills recorded in the coastal savanna (three crowned duikers *Sylvicapra grimmia* Linnaeus 1758 and three cane rats) were attended by twelve crows, four vultures and one kite. There was only one conflict, which saw a vulture drive a crow from the ground to a treed perch.

Discussion

The results of the analysis support the following points: (1) the primary factor for hooded vulture abundance was urban landuse, related to the abundance of organic waste and nearby vegetation density; (2) urban landuse and cover was slightly more important in this respect than rural landcover and vegetation density; (3) hooded vultures and their competitors generally, but not entirely, preferred urban to rural areas for feeding; (4) there was some competition between the main bird scavengers (5) and despite these dynamics, hooded vultures nested mainly in rural rather than urban areas. While some of these points have been briefly mentioned in the general, non-Ghanaian literature (Ferguson-Lees and Christie 2001), several neglected details have been documented in this study.

There were noted preferences for urban areas near to dense vegetation for roosting and nesting. Town size was found to be a significant attractant in forest areas near to cities, although the urban bird density was only weakly correlated with town size.

As such dynamics have been neglected in the literature, there is little understanding of this issue (Campbell 2009). The urban survey documented poorer sanitation in the markets, business areas and residential districts of small towns and villages than in the big cities, which appeared to be a factor compensating for the attractive power of larger cities. The main factor favouring larger towns was the presence of large abattoirs and butcheries, which attracted the largest flocks of vultures, and to a lesser extent pied crows and black kites. These have been described as significant attractors, but have not been compared with other sites such as markets (Madge and Burn 1994; Ferguson-Lees and Christie 2001; Obiri-Danso and Jones 2002).

Few studies have attempted to document the dawn and dusk behaviour of bird scavengers in urban contexts, although Ferguson-Lees and Christie (2001) point out that the hooded vulture starts foraging earlier than other vultures. Studies have documented the necessity of thermals for vulture soaring (Newton 1979; Houston 2001). However, the current study documented vultures, kites and crows, all soaring birds, flying into the cities from urban green spaces, peri-urban and rural forests at dawn, at low altitudes on their way to favoured feeding grounds (markets, abattoirs and rubbish dumps). Such behaviour was also documented at dusk, when markets were being cleaned and unsold meats were discarded. This flight behaviour gives important information on the mapping of bird movements in urban settings.

The results also examined what was a moderately aggressive relationship between hooded vultures and pied crows, and the more amenable relationship between hooded vultures and black kites. This is rarely mentioned in the literature, the focus instead being on the dominance of other vulture species over hooded vultures at feeding sites (Bannerman 1930; Newton 1979; Ferguson-Lees and Christie 2001). The pied crow was a significantly more active scavenger than the black kite, and it occurred in larger numbers. Interactions included the mobbing of flying vultures by crows, conflicts over scraps on the ground (despite the superior size of the vultures, only marginal dominance was attained in such situations) and slight dietary differences. The greater agility in the crows' feeding, where they played the role of the small, more agile scavenger, appeared to fulfil the role played by the hooded vultures where larger vultures were common (Butchart and Roche 2003).

Conclusion

The hooded vulture performs valuable functions in both rural and urban contexts. There must be more studies on this, with comparisons between areas with and without other vulture species. In this study, although the black kite and pied crow, and in a few cases feral dogs, were competitors for carrion and other waste products, and there was evidence of aggressive engagement between these species, the hooded vulture was the dominant urban and rural scavenger. This situation is comparatively rare in sub-Saharan Africa, as most other areas, including northern Ghana, fall within the overlapping ranges of larger vulture species. Southern Ghana therefore provides a good area to contrast, within comparative studies, the hooded vulture's behaviour with that which obtains in the different contexts of East Africa.

There must also be detailed studies comparing urban and rural areas for scavengers, which has also been neglected in the literature. This paper has highlighted the merit of examining bird behaviour in both urban and rural contexts.

From this approach, it could be seen that these were not separate behavioural contexts, but there were strong inter-linkages with varying adaptive behaviour from the hooded vultures, black kites and pied crows. The scaling of such comparative analyses to wider contexts will exponentially increase the complexities. This would be necessary, due to the need for understand of the increased rates of urbanisation, human population increase and habitat modification, which may impact on bird scavengers and other raptors in the southern continents.

References

Bannerman, D. A. 1930. *The Birds of Tropical West Africa*. 1. London: The Crown Agents of the Colonies.

Baker, M. 2001. *The Birds of Kitumbeine Forest Reserve: Using Geographic Information Systems to Outline Habitat Selectively in Forest Bird Communities*. New York: UNDP.

Bayne, E. and Hobson, K. 2000. Relative use of contiguous and fragmented boreal forest by Red Squirrels (*Tamiasciurus hudsonicus*). *Canadian Journal of Zoology*, 78: 359-365.

Bensen, P. C. 1999a. An evaluation of some techniques and problems associated with marking of Acciptrid Vultures. In: Boschoff, A. F., Anderson, M. D. and Borello, W. (eds.) *Proc. Vultures in the 21st Century – Vulture Research and Conservation in Southern Africa*. 148-150. Kimberley: Northern Cape Nature Conservation Service.

Benson, P. C. 1996b. Status of vultures in the Northern Province South Africa. In: Boschoff, A. F., Anderson, M. D. and Borello, W. (eds.) *Proc. Vultures in the 21st Century – Vulture Research and Conservation in Southern Africa*. Kimberley: Northern Cape Nature Conservation Service

Biggs, D. 2001. Observations of eagle congregations at a Red billed Quelea colony in the Kruger National Park. *Bird Numbers*. 10 (1): 7.

BirdLife International (2013a) Species factsheet: Necrosyrtes monachus. Downloaded from http://www.birdlife.org on 18/04/2013.

BirdLife International (2013b) Species factsheet: Gyps africanus. Downloaded from http://www.birdlife.org on 18/04/2013.

BirdLife International (2013c) Species factsheet: Trigonoceps occipitalis. Downloaded from http://www.birdlife.org on 18/04/2013.

BirdLife International (2013d) Species factsheet: Neophron percnopterus. Downloaded from http://www.birdlife.org on 18/04/2013

Brown, L. H., Urban, E. K. and Newman, K. 1982. *The Birds of Africa*. 1. London: Academic Press.

Campbell, M. 2009. Factors for the presence of avian scavengers in Accra and Kumasi. *Area* 41(9): 341 – 349.

CIA 2004. *The World Factbook*. URL: http://www.cia.gov/cia/publications/factbook /geos/gh.html

Cramp, S. and Perrins, C. M. 1977-1994. *Handbook of the Birds of Europe, the Middle East and Africa. The Birds of the Western Palearctic*. Oxford University Press, Oxford.

Dickson, K. B. and Benneh, G. 1995. *A New Geography of Ghana.* London: Longman.

Dowsett, R. J., Forbes-Watson, A. D. 1993. *Checklist of Birds of the Afrotropical and Malagasy regions*. Tauraco Press, Li.

Ferguson-Lees, J. and Christie, D. A. 2001. *Raptors of the World*. Christopher Helm, London.

Finn, M. 2004. URL: http://www.free-living.com/where%20to%20watch/ghanawtwb.

Ginn, P. J., McIlleron, W. G. and Milstein, P. le S. 1989 *The Complete Book of Southern African Birds*. Cape Town: Struik Winchester.

Government of Ghana 2000. *Census 2000*. Accra: Government of Ghana.

Houston, D. 2001. *Vultures and Condors*. Granton-on-Spy: Colin Baxter.

Iddi, K. N. 2004. *Sustainable Use of Birds in Ghana and Mitigation of Constraints to Conservation of Threatened Species*. London; Whitley Laing Foundation.

Madge, S. and Burn, H. 1994. *Crows and Jays*. London: Christopher Helm.

Mundy, P., Ledger, J. and Butchart, D. 1997. *The Vultures of Africa*. Academic Press, London.

Newton, I. 1979. *Population Ecology of Raptors*. Berkhamsted: T and A.D. Poyser.

Ntiamoah Baidu, Y. 1987. West African wildlife: a resource in jeopardy. In: *Unasylva* 156 - *Tropical Rain Forest Management: A Status Report*. Rome, FAO.

Obiri-Danso, K. and Jones, K. 2002. Campylobacters and faecal indicators in vultures at Kumasi abattoir, Ghana. *Abstract for poster at Society for Applied Microbiology summer conference on Pathogens in the Environment and Changing Ecosystems*, Nottingham, July. (Biomerieux Organon Teknika Poster Prize) 2nd

Sibley, C. G., Monroe, B. L. 1990. *Distribution and Taxonomy of Birds of the World*. Yale University Press, New Haven.

Stronach, N. 2003. Checklist of the birds of Selous Game Reserve. In: Baldus, R. D. and Siege, L. (Eds.) *Tanzania Wildlife Discussion Paper* 35: 1-35. Dar Es Salem: Wildlife Division, GTZ Wildlife Programme,

Tewksbury, J., Hejl, S. J. and Martin, T. E. 1998. Breeding productivity does not decline with increasing fragmentation in a western landscape. *Ecology* 79: 2890-2903.

Thiollay, J. M. 2006. Severe declines of large birds in the northern Sahel of West Africa: a long-term assessment. *Bird Conservation International* 16(4): 353-365.

Virani, M., Kendall, C., Njoroge, P., Thomsett, S. 2011. Major declines in the abundance of vultures and other scavenging raptors in and around the Masai Mara ecosystem, Kenya. *Biological Conservation* 144: 746-752.

Zimmerman, D. A., Turner, D. E. and Pearson, D. J. 1999. *The Birds of Kenya and Northern Tanzania*. Princeton University Press, Princeton.

Index

A

B

G

H

I

J

K

L

M

N

O

P

T

U

V

W

X

Y

Z